FRENCH ON SHIFTING GROUND

AMERICA'S
THIRD
COAST

Carl A. Brasseaux and Donald W. Davis, series editors

FRENCH
ON SHIFTING GROUND

Cultural and Coastal Erosion in South Louisiana

Nathalie Dajko

University Press of Mississippi / Jackson

This contribution has been supported with funding provided by the Louisiana Sea Grant College Program (LSG) under NOAA Award # NA14OAR4170099. Additional support is from the Louisiana Sea Grant Foundation. The funding support of LSG and NOAA is gratefully acknowledged, along with the matching support by LSU. Logo created by Louisiana Sea Grant College Program.

The University Press of Mississippi is the scholarly publishing agency of the Mississippi Institutions of Higher Learning: Alcorn State University, Delta State University, Jackson State University, Mississippi State University, Mississippi University for Women, Mississippi Valley State University, University of Mississippi, and University of Southern Mississippi.

www.upress.state.ms.us

Designed by Peter D. Halverson

The University Press of Mississippi is a member of the Association of University Presses.

First printing 2020

∞

LCCN 2020032215
ISBN 9781496830647 (hardback)
ISBN 9781496830937 (trade paperback)
ISBN 9781496830944 (epub single)
ISBN 9781496830951 (epub institutional)
ISBN 9781496830968 (pdf single)
ISBN 9781496830975 (pdf institutional)

British Library Cataloging-in-Publication Data available

CONTENTS

PREFACE

This book introduces readers to the lower coastal marshes of Lafourche and Terrebonne Parishes, Louisiana, where both the land and a language—Louisiana French—are in danger of disappearing. The book takes advantage of these unusual circumstances to examine the theoretical concept of place. It considers the close ties between the language, the land, and identity and demonstrates through various lines of inquiry that in Terrebonne-Lafourche, a place-based identity—referred to here as Bayou identity—underlies all other affiliations. I show that place is an abstraction, the representation of a community, mapped onto the physical landscape as well as the aural soundscape. Because places both reflect and form community and individual identity, they are inherently personal; people's attachment to both land and language consequently runs deep, and the loss of either is felt very keenly.

The tone of the book is at once scholarly and deeply personal (as was the research from which it is derived). This is intentional. I present data from a region in which the literal foundation on which people's homes and communities are built is washing away beneath their feet, a region in which the language of the oldest generation is lost on the ears of the young. This is a story about the construction of place via language, but it is also a story about deep personal loss. Consequently, I weave together linguistic data, observations culled from years of fieldwork and participant observation, and the words of the people who live there. Chapter 8, in which the story of the 1915 destruction of Leeville is presented almost exclusively in the words of the man who told me the story, is the most extreme example, but I have attempted as often as possible to include direct quotes with as much context as space and clarity permit. The combination of approaches allows for a scientific investigation into a phenomenon without losing sight of the fact that the phenomenon is inherently personal: the creation and loss of place via both land and language in a part of Louisiana that by now has become famous in the popular press.

This book is the culmination of a decade and a half of research on Louisiana's French varieties. My earliest fieldwork experiences (as a graduate student, part of a class at Tulane University) took me to the prairie parishes of Evangeline and Avoyelles. I followed that up by spending the summer of 2003 interviewing the last handfuls of speakers in peripheral areas: in and around Natchitoches, Bayou Lacombe (St. Tammany Parish), and notably the Buras-Diamond area in Plaquemines Parish, just downriver from New Orleans. In Plaquemines, I first encountered the indigenous population of Louisiana's coastal marshes as well as the cultural effects of coastal erosion. I learned much about the complexity of identity in francophone Louisiana and began thinking about the links between coastal erosion, language loss, and identity. That summer, I met people whose identity centered on being "French" (as opposed to either "Cajun" or "Creole," the most well-known terms used by and in reference to francophone Louisianans). I rode on a boat to interview people whose only access to their homes was via the water. I heard stories of flooding following hurricanes, of escaping into the marshes to ride out storms on fleets of boats tied together for safety, of driving cars up onto the levee in the hopes the water wouldn't rise that high and wash them away, of towns that no longer existed because they had been claimed by the sea. I also heard stories of being denied the right to speak one's native language and of choosing to speak to one's children in a different language and later deeply regretting that decision. Of words that were lost in a sea of new voices. "Oh, mon beau langage ... [Oh, my beautiful language ...]," an interviewee in East Pointe à la Hache sighed as he stared wistfully into the distance.

Three years later, in 2006, returning from a semester in New York state following a forced exile due to Hurricane Katrina, I witnessed firsthand the damage not only to my own city—my adopted hometown—but also to those places in which I had spent so much time sitting in the yard or around the kitchen table hearing stories of the ways the culture and the landscape had changed over the course of a lifetime. For three weeks, I worked with Save the Children to document living conditions in internally displaced people (IDP) camps across Mississippi, Alabama, and Louisiana. At the time, I was living in a tiny apartment in my now-crowded neighborhood in New Orleans, only a block away from the flooding. I remember vividly feeling incredibly lucky by comparison with friends who had lost everything and with those I met in IDP camps, many of whom were living in RVs filled beyond capacity in remote state parks or in airport parking lots. One week, we were housed by Save the Children just a few blocks inland on the Mississippi Gulf coast; nearby homes had been reduced to piles of rubble or stairs leading to empty concrete slabs. In Plaquemines Parish, where I had spent so many hours sharing some small part of people's lives, only one building remained standing; everything else had been entirely obliterated by the storm, the residents scattered or living in campers. Whenever I met residents from coastal Louisiana housed far inland, they immediately commented on the absence of the French language in their current setting. That spring and summer, I also made my first trips to Terrebonne and Lafourche Parishes, conducting preliminary interviews that would form the basis of my dissertation work on variation in Louisiana French. The area had been hit hard by Hurricane Rita, which struck a few months after Katrina, and relief groups, most notably Mennonite Disaster Services, still maintained a presence. Again and again, my interviews turned to the topic of the dramatic changes to the physical landscape. Again and again, people lamented the disappearance of the language.

I first went to the lower Lafourche basin with the goal of documenting variation in the French spoken there to determine whether it patterned along ethnic and geographic lines. When I presented my finished dissertation to some of the people I interviewed, one person commented, "You know what you should really do? You should tell the story of what this place is *really* like." This book is my attempt to at least begin to do so. It is my attempt to define what "this place" is and to demonstrate the important role language plays in the construction of place. Place is inherently personal; for that reason, its loss is felt acutely, and everyone should care about displacement, about language loss, and specifically about the loss of Louisiana's coastal marshes.

A NOTE ON THE CONVENTIONS USED IN THIS BOOK

In this book I present a linguistic examination of French in Louisiana and the symbolic role it plays. This involves the occasional presentation of excerpts from interviews conducted with participants. Normally in a linguistic study, one uses nonstandard punctuation in presenting excerpts of speech in an endeavor to present as accurately as possible the realities of speech: sentences are hard to identify because people often backtrack or stumble at the beginning of an utterance, people pause (or don't) in unexpected places, their speech often overlaps, and so on. However, doing so would render the passages more difficult to read. I have opted instead for a hybrid system: I have used standard punctuation but retained as much of the information in the original as possible. This means that question marks are used to indicate questions even if rising intonation was not used in their asking. I have divided the utterances into readable sentences. On the other hand, I have not deleted instances of false starts and have been as faithful as possible to the recordings. As such, the following conventions are used:

- Speech that was unclear but for which I had a good guess as to what was said is put in parentheses: (I walked down the street) indicates that I believe what was said is "I walked down the street," but am not certain.
- Speech that I could not understand at all is indicated with an X for each syllable: XXX thus indicates three unclear syllables.
- Sounds that were drawn out are indicated by a colon. For example, "u:m" indicates that someone pronounced *um* with a lengthened vowel.

I have provided translations for all French utterances. In such cases, the text appears in two columns: French on the left, and English on the right.

It is common practice in anthropology and linguistics to use pseudonyms rather than participants' real names. I like to recognize people's contribution to my research whenever possible, however—it is *their* language and *their* culture, after all—and almost all my participants elected to be identified by name in publications that might result from our interviews. Moreover, some of the dialogue presented here was in fact filmed; a short movie featuring it appears on YouTube and at my website (www.tulane.edu/~ndajko), and I have been doing recovery work on a longer film as well. It would be absurd, consequently, to try to hide some of my participants' identities behind pseudonyms. In any case, I have not provided names for every short speech excerpt; for short excerpts I have kept my transcription conventions, by which researchers are labeled R and interviewees L (for the French *locuteur* [speaker]), with numbers following to distinguish them from each other. So, for example, R1 is the first researcher, R2 the second, L1 the first interviewee, L2 the second, and so on. I do provide names for people profiled at any length. Some of these names are real, and others are pseudonyms, following the wishes of the interviewees.

ACKNOWLEDGMENTS

As is always the case, this book could not have been written without the support of a good number of people. For comments on earlier drafts, I thank Shana Walton, Mike Picone, Allison Truitt, Tom Klingler (who provided the title), and several anonymous reviewers, each of whom offered a critical eye and constructive suggestions that helped me bring data from a broad array of sources into the narrative. If this text is still in some way lacking, the blame falls squarely on me.

In the field, I was rarely alone. First and foremost among my assistants were Rocky McKeon, who taught me about local English, French, and the folk history of the area, and Roland Cheramie, who taught me to play Cajun music. Carrie Johnson accompanied me on my first summer of fieldwork, taking excellent notes; Katie Carmichael joined me later and let me tag along on some of her trips. Darcie Blainey did likewise; both her insights into the region and those of Katie have helped form my ideas. Darcie Blainey, Matt Crossland, Audrey Fort, Amber French, Wilson Goss, Caroline Hinrichs, Carrie Johnson, Rebecca Powers, Robert Pulwer, Grace Schipps, Alexander Schneider, and Sara Snider were the best fieldwork team I could ever have hoped for in the spring of 2008. Zach Hebert and Shane Lief helped with data collection for the map task presented in chapter 7; their input as native Louisianans was especially important. Sarah Reynolds helped me hunt down terms in indigenous languages.

Katherine Bell, Nathan Wendte, Maxime Lamoureux St.-Hilaire, Rocky McKeon, and Amber French helped with transcription, and Roland Cheramie, Alces Adams, and Christine Verdin lent their native-speaker ears to verify transcripts.

Financial support came from the Louisiana Board of Regents; the Jacobs Research Fund Kinkade Grant from the Whatcom Museum in Bellingham, Washington; and the National Science Foundation (NSF #0745971).

Judy Soniat at the Terrebonne Parish Library very enthusiastically helped me conduct background research, collecting sources for me and pointing me in the direction of things I would never have known to consider. Likewise, the staff at Special Collections at the Howard-Tilton Memorial Library at Tulane helped me with background research. My mother, Marieke Dajko, helped with genealogical research throughout, going above and beyond to provide me with more data than I could ever use.

The rest of my family, in particular Tom Klingler, who read multiple drafts and made sure the house was never out of milk, has also been a source of continual support, both moral and material, as have friends who are like family. JoAnn Burak stepped in to watch our son one evening in Paris as I worked all day and night to finish writing and his father had work obligations.

Working with the University Press of Mississippi has been an incredibly rewarding experience. In particular, I thank Craig Gill for several years of support through two different books. Ellen Goldlust was wonderful yet again at making my words sound better.

Most importantly, however, I thank the people of Terrebonne and La-fourche Parishes. I owe an enormous debt of gratitude to the roughly 250 people who invited me into their homes and shared their food, their language, and their knowledge with me as well as to the dozens more who participated in surveys or shared their insights over meals or at public gatherings. I especially thank Chuckie Verdin, Melissa Billiot, Patty Ferguson, Randy Verdun, Albert Naquin, Glen Pitre, Emilia Pitre, Billy Pitre, Bobby Samanie, Jake and Esther Billiot, Donald and Theresa Dardar, Arline Naquin, Christine Verdin, Geneva LeBoeuf, Roland Cheramie, Rhonda and Jody Billiot, Roland (Roach) Cheramie, Alces Adams, Toot Naquin, Megan Neil, Lynnell and Donald Gaspard, and Cheril Edens, who made me feel like I was part of the family. I'm not sure I can ever repay them, though I will certainly try.

FRENCH ON SHIFTING GROUND

Chapter 1

SHIFTING LAND, SHIFTING LANGUAGE

"Quand on a vu l'eau monter, on a se dépêché vite, et on a menu à la place de votage [When we saw the water rising, we hurried and came to the voting place]," Toot Naquin began, cigarette in hand. Outside, neighbors and relatives moved what was left of waterlogged possessions, drywall, and bags of insulation to the edges of their yards. Heaps of debris piled up at the roadside, waiting to be removed. As she talked, Toot periodically looked through the window she was sitting next to, keeping an eye on the cleanup. This was only the second time in her seventy-eight years that she had seen the storm surge from a hurricane push its way far enough up Bayou Pointe au Chien to flood her land; the first had been three years earlier, in 2005, when Hurricane Rita had left three inches of water in the house. Upon seeing the fast-rising floodwaters, Toot and her family—her daughter, son-in-law, granddaughter, and two great-grandchildren—had crossed the street to what had once been a dance hall but now functioned as a polling place during elections; it was a few feet higher than the house.

"L'eau a monté vite [The water rose quickly]," added her son-in-law, taking a break from cutting up debris to join the interview.

"Ouais, elle a monté vite," she agreed. "Eusse m'a emmené icitte dedans un char, et faullait que je monte avec le walker [They brought me here in a car, and I had to use the walker to come inside]." She paused.

"Dans l'eau [In the water]," her son-in-law reminded her.

"Bon pour me noyer! [Enough to drown me!]," she finished, laughing dryly.

SHIFTING LANGUAGE

Toot's one-story brick house halfway down Highway 665 in the town of Pointe aux Chênes was built on land her father had bought years earlier because it was high above any point on the bayou that had ever flooded.[1]

3

Table 1.1 Census Counts of French Speakers, 1990–2013					
1990 Census		**2000 Census**		**2009–13 ACS Estimate**	
French	227,717	French	179,745	French	101,330
Cajun	27,613	Cajun	14,355	Cajun	18,470
Creole	6,310	Creole	4,470	Creole	6,706
		Patois	215	Patois	130
Total	261,640	Total	198,795	Total	126,636

Source: 1990 and 2000 U.S. Censuses; U.S. Census 2013

She spoke to us in her native French, a language she had spoken exclusively before beginning school at the age of six and that she still spoke with anyone who also spoke it. Those people, fewer in number and greater in age with every passing year, included her son-in-law, a native of neighboring Lafourche Parish.[2] The rest of the family had varying degrees of familiarity with the language. I was there that day with my frequent cointerviewer and local resident (from the nearby town of Robinson Canal), Rocky McKeon, to document on video the destruction to lower Terrebonne and Lafourche Parishes following two hurricanes that had hit earlier that month. We spoke to Toot in French in part because it was the language we had always spoken with her but also because we wanted to use the video we were making to stress the cultural loss that went hand in hand with the destruction of the land that we were witnessing.

French has been spoken in Louisiana for three hundred years, but today it is rapidly giving way to English. Table 1.1 shows the US census numbers for 1990 and 2000 and the 2009–13 American Community Survey estimate. The number of (self-declared) francophones dropped by roughly 62,000 over the first ten-year period and by roughly 72,000 over the ensuing decade, for a total decrease of about 50 percent over a twenty-year period. While census counts are problematic and the actual number of francophones can consequently only be approximated, these numbers are comparable to each other (since all of the data reflect the same problems) and provide a good indication of the rate of attrition.[3]

Terrebonne and Lafourche Parishes are among the most francophone in the state today. Figure 1.1 shows the percentage of francophones in Acadiana, the twenty-two-parish cultural region in which most French speakers live. Lafourche is the fourth-most-francophone parish, at 19.2 percent, a rate that surpasses that of Lafayette, often seen as the heart of French Louisiana and the home of the state's Council for the Development of French in Louisiana

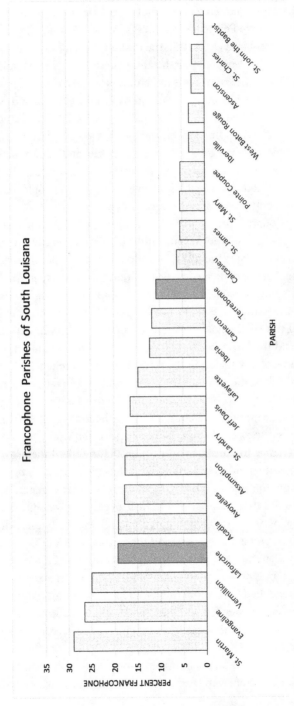

Figure 1.1. Percentage of French speakers per parish. Image created by the author. Data from U.S. Census 2000.

(CODOFIL), which is tasked with promoting the language in the state. Terrebonne Parish, where Pointe aux Chênes is located, is not quite so heavily francophone but still falls well within the midrange of francophone parishes at 10.66 percent. What overall parish rates cannot show, however, is localized concentrations. The rural areas in Lafourche and especially Terrebonne have rates much higher than the overall parish rate. In the town of Golden Meadow near the southern extreme of Lafourche Parish, for example, the rate is 38.8 percent. Given that most francophones are elderly, this is a dense concentration indeed. Even today, Terrebonne has a handful of monolingual French speakers possessing only receptive competence in English (that is, they understand but do not speak). I interviewed six such people during my initial fieldwork and was aware of several others.

Despite this relative robustness of the French language, most speakers are over the age of sixty, if not seventy. Indeed, I have met only a handful of speakers born after 1970 (generally the children or grandchildren of residents who were particularly dedicated to using French exclusively in the home), though many younger people have receptive competence or, as is most common among the youngest generations, can repeat a few stock phrases and expressions. Even so, the concentration of fluent speakers in Terrebonne-Lafourche—and the lower reaches of the bayous in particular—has one of the best chances of any region in the state of reviving French.

In these parishes, the generations born after World War II have seen a precipitous drop in the use of French, as is true across Louisiana. The decline likely has roots in a number of factors, including the postwar economic boom that favored the use of English as the language of the workplace (Bernard 2003), a 1921 change to the state constitution that required the use of English in the classroom and that compounded the effects of 1916's mandatory schooling regulation, and the fact that members of the first generation that had been required to learn English at school began to have children in the mid- to late 1940s and had the option not to speak French to their children. (See Bernard 2003 for a full discussion of the factors contributing to the shift to English in the twentieth century.) The majority of those I interviewed had learned English only when they began school at the age of six; for the most part, their parents had been monolingual French speakers. Many residents remember the shame they felt when entering school speaking only French and consequently refused to speak it with their own children, a decision they generally regret today. Their children and grandchildren are often sorry that they have not learned the language fluently and that they thus cannot pass it on to their own children; teenagers seem less concerned, but conversations with young adults, in particular those who are parents, suggest that this may well be an age-graded phenomenon and that regret will set in as they get older. Many people in these younger generations, particularly those only slightly younger than the youngest fluent speakers, are semispeakers, meaning they can manipulate the language to some degree but do so imperfectly and with difficulty. The youngest generations often know a few stock words and phrases. While some non-speakers can understand the language, most cannot.

As a result of the high concentration of elderly speakers, French can still be heard in public places like restaurants and stores when older residents are present. That said, the language is most often heard in private settings, between spouses or friends. When older residents attend informal community gatherings, French is also used despite the presence of younger non-francophones. Out of deference to the younger speakers, however, public meetings to discuss important business take place in English; though there are still a handful of French monolinguals, most fluent francophones are fully bilingual. Because English was the only language of instruction in schools and French was rarely taught even as a second language, nearly everyone is illiterate in French (though of course they can read in English), even if they speak it fluently. Several people do remember parents or grandparents who could read in French, however; it was never made clear to me why the ability to read in French had not been passed on, though presumably it is linked to the shame felt by many francophones and the emergence of English as the only language of the workplace, the school, and in many people's minds the future.

French is still heard in limited public places, however. On one of my first trips to lower Terrebonne Parish, I arrived on the evening that the annual "French mass" (actually the French Rally), was being held at Live Oak Baptist Church in lower Pointe aux Chênes. Lafourche radio station KLRZ (The Rajun' Cajun), based in Larose and heard in New Orleans, seventy miles away, had a weekly Sunday morning broadcast, hosted by Gloria Fonseca, who speaks French for up to 50 percent of her time on-air. The language is also still used in workplaces dominated by francophones, particularly the community of shrimpers who ply the bayous and lakes in their skiffs. I accompanied several shrimp fishing excursions over the years; in the warmer months, shrimping is done at night, when the temperature drops slightly. Over the rumbling of the diesel engine and the squeaking of the winches and pulleys as the nets are brought in and sent back out was the sound of the shortwave radio, crackling in French as the (mostly male) fishing crew talked to each other through the night. The disappearance of French is not a punctuated event; it is taking place over generations, only recently gaining in speed. At the same time, its speakers sit on land that is undergoing the same process.

SHIFTING LAND

That Louisiana's coastal marshes are disappearing is a long-established fact. The land has seen such rapid and drastic changes that even people in their twenties and thirties can talk at length about how the landscape has changed since they were children. For several decades, a steady stream of ever-louder warnings has come from official agencies and the press, who generally present the pace of loss in terms of how many football fields of land are lost every year, month, day, or even hour (see, e.g., Kennedy 2017; AJ+ 2015, both of which cite one football field per hour). It has become nearly impossible to keep up with the volume of articles and news reports documenting what has become a dire situation in the coastal marshes that form Louisiana's Gulf Coast. I found seven such reports in various popular media (print, broadcast, and online) in the first four months of 2017 alone, and their publication continues apace. An article in the January 2017 issue of *The Lens*, an online investigative news publication based in New Orleans, as part of a series focusing on coastal land loss summarizes the growing alarm. Author Bob Marshall (2017) reports on the newly released Comprehensive Master Plan for a Sustainable Coast issued by the Coastal Protection and Restoration Authority of Louisiana, which reveals that the worst-case scenario in the state's 2012 Master Plan had become the best-case scenario just five years later: tens of thousands of homes in coastal Louisiana, in particular the lower bayous, will need to be bought out or elevated. Meanwhile, Hauer, Evans, and Mishra (2016:693) predict that millions of people in the continental United States will be displaced by rising sea levels over the next century, with Terrebonne and Lafourche Parishes among the most severely affected counties in the nation.

On April 18, 2017, Louisiana governor Jon Bel Edwards declared a state of emergency over the disappearing coastline and asked the federal government to do the same (*State of Emergency* 2017). Both Hauer (2017) and the governor's declaration of emergency note the national importance of the eroding wetlands. The declaration of emergency cites, among other reasons, the fact that five of the nation's fifteen major ports are located in Louisiana, that they handle 60 percent of the country's grain commerce, and that Louisiana's pipelines handle 90 percent of the country's offshore energy production. Hauer points out that the influx of people migrating from the coasts will have real effects on the areas they choose to move to, further broadening the consequences of coastal erosion. Louisiana's disappearing coast is decidedly *not* simply a local problem.

The destruction of the land is in constant evidence: dead trees, killed by saltwater intrusion into what used to be solid ground, poke out of the open marsh. Even the most casual visitor will notice this indicator, but for those who have spent a lifetime in the area, it is a constant, jarring reminder, as the oldest have memories of a time when the area was thickly forested, and even young people can detail physical changes that they have witnessed. Today, open water has replaced land where cattle grazed and gardens grew. Seawater often covers roadways in the lower reaches of the parishes when the tide is high or a stiff wind blows from the south. In lower Lafourche Parish, telephone poles stand half submerged by seawater, a persistent reminder of the ongoing change. In putting together a documentary video using my footage, I came across a French news report filmed in the late 1960s on the Île à Jean Charles (Failevic 1969).[4] A house flanked by trees appears on the screen; Rocky, my fieldwork partner, recognized it as a house we had passed many times on the Island. Now abandoned (the residents moved to a newer house next door and simply left the old one to fall apart), the house sits in on a narrow, treeless ridge only feet from open marshland. In a similar vein, while interviewing an elderly resident of the island I noticed on the shelf behind him a photo of a young woman standing on the porch of a house with thick trees all around. I was stunned to learn that the photo was of his wife, taken decades ago in front of the same house we were sitting in, though it was now fifteen feet above the ground and surrounded by miles of grass and water with only a few scattered trees, many of them dead or dying.

In June 2009, I took a boat tour around modern Leeville to film the state of the marsh. The tour guide was initially reluctant to take my money, saying that it was more important to him that I fully understand what was going on. We sped past half-submerged telephone poles before slowly circling around a small cemetery—or rather, what was left of it: a handful of graves, literally crumbling into the water. I was briefly alarmed when we approached a large patch of grass at full speed, but to my surprise, we sailed right over it: the grass was growing on land that was now fully underwater, and the flat-bottomed boat easily passed over it. As we passed over patch after patch of the same, my guide explained that only a few months earlier, this had all been dry ground. I had seen this before in more abstract form: the shrimp boat I sometimes accompanied on trawling trips navigated with a GPS map that often indicated that we were on land when in fact the boat was in open water. But green pixelated patches on a screen are far less arresting than driving a boat straight into what looks like solid land.

The next day, an excursion to film the eroding bayous likewise provided a graphic illustration of the change the area has undergone. Having heard from multiple interviewees in Pointe aux Chênes that they had grown up miles below the current settlement, I asked a friend, also one of my interviewees, to take me to the old settlement. A member of the Pointe-au-Chien Indian Tribe, he had been raised in the town and had begun his shrimping career at the age of thirteen; he was very familiar with the changes affecting the marsh.[5] On the way down I took video footage of narrow strips of land marked by dead and dying trees.

"I don't know about this!" my guide said, as he cut the engine upon banking at our destination, and I turned around to see him flailing at the air.

I immediately discovered why as I stood up to get out of the boat: deerflies—insects with a mean bite that thrive in swampy environments—so thick we couldn't get out. We pulled the boat out and banked at an alternative spot, and while we were able to walk around a bit there, my camera tells the story of the miserable few minutes it was: shots of our feet, of the ground, and of the air as the camera is flung about as I desperately swat at deerflies. My friend's attempt to explain where we are is punctuated by the sounds of us slapping and cursing at the flies. We made a side trip on the way back to see Fala and l'Esquine, two tiny settlements—only a few houses each—hidden deep in the marshland, entirely unreachable except by boat and on their way to disappearing. Along the way, my friend talked about how in his youth, forty years earlier, the areas we were passing through had been so thick with trees that it stayed relatively cool even during the hottest summer months; today, though a few trees remained, we sat directly in the hot sun. The area is entirely uninhabitable. The cross that marked the cemetery where the ancestors of the modern indigenous population are interred was a sober reminder of just how much people are giving up when they are forced to abandon an area. The connection to the land is literal: family is buried there. One is leaving behind something more than a symbolic attachment to a place.

In perhaps the most comical but most disturbing scene, following up on an interest some archaeologist friends of mine had expressed in studying the Indian mounds that lie below the current settlements, I took my satellite map down the bayou to ask about the mounds and to see if the community would be interested in an archaeological study. A group of shrimpers attempted to circle the areas on the map where the mounds were located, but the men began arguing because they couldn't recognize landmarks on the map; it showed land where there should be water.

"This map is way out of date," they complained repeatedly.

The map was three years old.

The causes of coastal land loss include levees on the Mississippi River that prevent new sediment from being deposited in the delta, dams upriver that prevent much of the sediment from reaching the lower delta, canals cut by oil companies that weaken the land and more readily allow erosion and saltwater intrusion, and climate change resulting in sea level rise. Climate change is also expected to bring stronger hurricanes, which, in combination with sea level rise, will push storm surges further inland (Geophysical Fluid Dynamics Laboratory 2018). Hurricane Katrina brought the issue to national prominence in 2005, but though they suffered some damage, lower Lafourche and Terrebonne Parishes were hit harder by Rita later that year and then twice in the space of two weeks by Hurricanes Gustav and Ike in 2008. Though less intense, these storms were more destructive to the lower bayous than was Katrina (which only brought strong winds) because the storm surges inundated the area with several feet of water and left it covered in mud and debris when the water receded. In recent years, storms have brought flooding to areas that had never previously experienced it. Today most houses in the lowest reaches of the bayous are elevated ten or more feet. This is a recent development; photos from the mid-twentieth century show no more elevation than that which is common in South Louisiana (a few feet to allow air to cool homes) if any. Toot, for example, pointed out that she had built her home on a concrete slab precisely because the area had no history of flooding. Each year, the elevation of houses moves further up the bayou.

LANGUAGE AND PLACE

As we sat in the old dance hall with Toot Naquin in September 2008, watching the cleanup following Hurricanes Gustav and Ike, we discussed the changes taking place to the land using a language that was equally in danger of disappearing. The town had been without power for nearly a month. Toot's house had received a foot of water, and in addition to needing to repair the lower portion of the walls in all the rooms and replace the floor, the family had lost the washer and dryer, several large pieces of furniture, and the central air-conditioning (a particularly tough loss in Louisiana's hot, humid climate). For the past month, they had been doing laundry by hand and hanging it on a line to dry. Toot, who had grown up in a time before electricity had come to the lower bayou, was not relishing this virtual trip to the past.

"This is a hell of a life," she told us, switching to English with no trace of her usual underlying humor.

"Je crois pas que je vas voter encore [I don't think I'm going to vote again]," she added through exhaled smoke: "Tu votes pour le bougre, et pour qui?! Mon, je vois pas que ça peut arranger arien, moi. C'est honteux [You vote for the guy, and for what?! Me, I don't see that it can fix anything, me. It's shameful]."

But her strongest words were uttered in response to our wondering if she would be moving up the bayou now that it was flooding here. She stopped smoking and stared directly into the camera as she declared that she would never leave: "J'étais énée sus le bayou, j'étais élevée sus le bayou, et je vas mourir sus le bayou! Et pas d'ouragan qui va me prendre! [I was born on the bayou, I was raised on the bayou, and I'm going to die on the bayou! No hurricane is going to move me!]."[6]

Toot's attachment to the bayou makes it clear that the disappearance of the land in Terrebonne-Lafourche is not simply the erosion of earth: it is earth that matters to people at a deep psychological level—a *place*—that is vanishing beneath their feet. In this book I explore the theoretical construct of place and the role that language (in this case, French) plays in its construction. In doing so I describe the process Picone (1997b:137) predicted two decades ago now coming to fruition: the loss of the land precipitating language loss. The lower coastal marshes of Louisiana may seem an unlikely choice for the study of place given the disappearance of the land; however, the simultaneous threat to land and language in fact throws the relationship between personal identity, language, and place into sharp relief and reveals the ways in which place is constructed both physically and aurally. Place-based identity—in this case, a connection to the bayou—underlies other identities. These layers of identity are accessed via the study of language. Places are community boundaries mapped onto the landscape, as other scholars have shown; I demonstrate that they are projected in parallel ways onto the *soundscape* (Schafer 1993) as well. The loss of language is as deeply felt as the loss of land, in part because place is being lost in both cases, and place is inherently personal.

The examination of place has been undertaken by scholars in many different fields, from the social sciences (such as anthropology, sociology, geography, history) to the humanities (literature) to the sciences (psychology) and beyond (e.g., architecture). Considering the range of scholars that have studied place, Gruenewald (2003:620) points out that it has no single definition. Still, Tuck and McKenzie (2015), conducting a cross-disciplinary synthesis of definitions, identify several common themes: place is subjective, it is constructed, it is abstract. Borrowing words from Tim Ingold (2008), they note that places "do not exist so much as they occur" (Tuck and McKenzie 2015:35). A *place* is distinguished in social science literature from *space* by its cultural foundation. *Space* can be roughly equated with physical landscape, the physical foundation onto which *place* is mapped. While *space* lacks boundaries and landmarks—it is essentially a blank slate, full of potential (and inviting for that reason)—*place* is familiar. *Places* are bounded spaces, with those boundaries marked by features that stand out from the background due to their cultural (and often also physical) prominence—that is, landmarks. And they, in turn, are locations at which events of importance to a community took place. *Place* is space possessed (Jarman 1993:126), lived in (Tuan 1975:164), and cared for (Schreyer 2008). As Low (1994:66) explains, a *place* is "space made culturally meaningful," "not just a setting for behavior but an integral part of social interaction and cultural processes."

Basso (1996b) further explains that places do not occur incidentally; rather, they are created by the stories people tell, using landmarks to ground their narratives. The limits of a place, in other words, are identified by the furthest extent of landmarks that have associated stories. Tuck and McKenzie (2015:65) likewise stress the importance of storytelling in place-making and the concomitant possession or stewardship of the spaces they bound, citing by way of example a dispute presented in Chamberlin (2001:127) between Gitxsan representatives and Canadian government officials regarding sovereignty over a forest that culminated in a Gitxsan elder asking pointedly, "If this is your land, where are your stories?"

Via storytelling, place-making is "retrospective world-building" (Basso 1996b:5). In telling the stories of the past, we create the modern place. Maldonado (2014:193–94) further elaborates that collective memory is selective—the stories that are told and retold shape the community's vision of the past and create the link between the people and the land. Such stories may impart community traditions and knowledge, hence Basso's title, which is a quote from an Apache participant in his research: *Wisdom Sits in Places*. Scholars of education such as Gruenewald (2003:620) emphasize the reciprocal nature of place-making and person-making: "As centers of experience, places *teach* us about how the world works and how our lives fit into the spaces we occupy.

Further, places *make* us: As occupants of particular places with particular attributes, our identity and our possibilities are shaped."

Tuan (1991:686) observes that the next step, naming a place, is "the creative power to call something into being, to render the invisible visible, to impart a certain character to things." Though Tuck and McKenzie (2015:13–14, 19) stress the inherently fuzzy nature of place boundaries and their temporal instability, to name a place is to create it. Moreover, as the Gitxsan example shows, place-making, including naming, is an act of taking possession or at least stewardship of a space. These are the spots at which important events occurred, and these events, when referenced or retold, continue to shape who we are today; they are, therefore, *our* spots to care for, to protect, and to exploit. Places and their names are consequently very personal, as they tell the story of a community and become intrinsically linked with the identities of its members. This is why places inspire loyalty and attachment (Tuan 1975:159). In Basso's (1996b:7) words, "What people make of their places is closely connected to what they make of themselves as members of society and inhabitants of the earth."

By focusing on language, I pick up where previous reports on the erosion of the lower coastal marshes of southeast Louisiana left off. In two books, Mike Tidwell (2004, 2006), who first came to the area with the intention of writing a simple travel report, describes the disappearance of the land supporting the culture he was sent to document and warns of the dangers of climate change. His first book describes the rapid pace of coastal erosion in Lafourche Parish and provides a vivid description of local culture. In 2006, he followed with an update, focusing heavily on New Orleans and Hurricane Katrina and discussing the underlying causes of land loss and the potential action to be taken to mitigate them. Both works briefly note the presence of French and of French-influenced English. McQuaid and Schleifstein (2006) document Katrina's destruction of New Orleans and consider its causes and the dangers that lie ahead using the Île à Jean Charles in Terrebonne Parish as a foreshadowing of what is to come. Writing four years later, Burley (2010) provides a phenomenological look at identity as it is affected by current land loss across coastal Louisiana. Based on hundreds of interviews conducted, he concludes that the erosion has resulted in an identity based in fragility. Most relevant to my work, Maldonado's (2014) dissertation is a study of community response to rapid land loss in three of the four Indian communities examined in this book (those located in Terrebonne Parish). Based on interviews both with people who have moved away and those who have chosen to stay, she concludes that the "vast and rapid land loss [has] created a sense of dislocation even for people who [have] stayed" (197). Her interviewees exhibited feelings of *solastalgia*, defined as a sense of homesickness people get without leaving home due to the changes to the environment that bring "a sense of dislocation and alienation from their subsistence-based activities, local, traditional knowledge, and memories" (212, citing Connor et al. 2004:55). Here I show that this dislocation and alienation result as much from the loss of language as from the loss of land.

METHODOLOGY AND OVERVIEW OF THE BOOK

In 2006–7 I conducted around 150 interviews in Terrebonne and Lafourche Parishes; that work was supplemented by approximately 100 interviews I conducted along with a class of students in Lafourche Parish in the spring of 2008. While conducting interviews, I spent as much time in the community as possible, and I continue to return there, though with less frequency as the years have gone by. I have participated in local festivals and celebrations and in local lifeways (I gained a small amount of notoriety by serving as a deckhand on a shrimp boat and because I had never before seen a gun up close or drunk rainwater from a cistern), attended community meetings (those of the American Indian communities and those of the Cajun French Music Association in Houma in addition to informal meetings and community get-togethers), and helped to clean up after Hurricanes Gustav and Ike and the 2010 BP oil spill.[7] I shared meals and learned how to fry crabs in a pot. I learned to play a few Cajun songs on the guitar and played with musicians from Golden Meadow. Following Gustav and Ike, Rocky and I videotaped the effects of the disaster and interviewed local residents and policymakers as we helped with the cleanup; I later also interviewed Kerry St. Pe, director of the Barataria-Terrebonne National Estuary Program, regarding the reasons for coastal erosion and possible solutions. In 2013, I hosted a panel at the New Orleans Jazz and Heritage Festival that brought together members of the Biloxi-Chitimacha Confederation of Muskogees from Grand Caillou, the United Houma Nation, and the S'Kallum (from the Pacific Northwest near Seattle) to discuss the effects of environmental change and displacement on community identity.

At the time, my direct goal was to document language variation linked to geographic origin and ethnic affiliation; thus, my linguistic interviews consisted of two parts: a translation exercise geared at identifying and documenting variation and a conversational section, conducted in French, lasting anywhere from twenty minutes to several hours (though generally about an hour) during which I gathered basic demographic information (age, gender, place of birth, ancestry, ethnicity) and discussed whatever themes interested my interviewees. Nearly every interview included a discussion of the physical changes to the environment. Nearly every interview also contained discussion of the loss of the language, with several interviewees explicitly connecting land and language, land and identity, or language and identity.

Following a description of the land and settlement history (chapter 2) and a brief history of the development of the French language in Louisiana (chapter 3), I turn to a discussion of the local dialect, the ways it is deployed to express ethnic and subregional identity within the parish, and the strong connection residents feel toward it (chapter 4). I then show, via perceptual work, that these documented and often recognized differences are subordinated to a general Bayou identity that is universal across the region and underlies other divisions and that this relationship is revealed via discussions of language (chapter 5). I then confirm this via an examination of a dispute over place-naming that superficially appears to be an ethnic dispute over stewardship but is in fact an argument over the characterization of the same place (chapter 6). Next, I illustrate that aural place is strong enough to survive even for those who no longer speak the language (chapter 7) and consider the ways in which physical and aural place are analogous via an examination of the stories told in the construction of place (chapter 8). These stories invoke both land and language, and people speak of the disappearance of land and of the language in parallel ways, demonstrating that place is both physical and aural space made meaningful. Language is the vehicle through which the stories used in physical place-making are told, but it is also symbolic in its own right.

In a critical response to several archaeological articles on the effects of migration on identity, Shepherd (2001:349) notes that according to the United Nations High Commissioner for Refugees, 50 million people had been displaced across both internal and international boundaries, at the beginning of the twenty-first century. Less than two decades later, that number stood at 70.8 million, an "unprecedented" figure that is the highest on record (United Nations High Commissioner for Refugees 2019). Since the publication of Shepherd's article, the world has seen the movement of millions of people as a result of drought, famine, and war. The United States has seen widespread devastation affecting vast swaths of its coastline including a major metropolitan area (New Orleans) and the coasts of Mississippi and Alabama following Hurricane Katrina in 2005 and the concomitant dispersal (in many cases permanent) of citizens around the region and the nation (Martinez, Eads, and Groskopf 2015). The 2017 Atlantic hurricane season was particularly severe. In August, Hurricane Harvey brought never-before-seen flooding to Houston and killed ninety people. Just a few weeks later, Irma devastated the islands of St. Martin and Barbuda, killing more than a hundred people and leaving over a thousand homeless. Barbuda was struck again only a week later by José, necessitating the evacuation of all of the island's population. At the end of September, Hurricanes Irma and Maria laid waste to Puerto Rico and the surrounding islands and resulted in the deaths of around three thousand people. Eight months later, in a striking reminder of the events that transpired in New Orleans following Katrina, the island had still not recovered, and thousands of people remained without electricity.

Shepherd (2001:349) notes that "the figure of the exile and the wanderer might appropriately stand as the iconic figure of late modernity." With predictions that millions of people in the United States alone will be displaced by sea level rise as a result of climate change (Hauer, Evans, and Mishra 2016; Hauer 2017), we can only expect migration, especially involuntary migration, to accelerate. As people become ever more mobile, we will need to understand the mechanisms of place creation and attachment if we hope to understand how displacement will affect both those who move and those they join in new places—whether those places are located within a few hours' drive of the original home or whether they are in an entirely different country.

Chapter 2

THE LAND AND ITS PEOPLE

Description of the Lower Lafourche Basin

The physical foundation of the place under study here, the Lafourche basin
is part of the intricate system of waterways and swamplands that protect the
American mainland and that are being taken over by the Gulf of Mexico with
increasing rapidity. In fact, much of the area included in current surveys of
towns and cities is actually waterlogged wetland or lake. Settlement tends to
follow the waterways in narrow strips that flank the banks. The lower basin,[1]
comprising Terrebonne and Lafourche Parishes (figure 2.1), encompasses
the delta of Bayou Lafourche (itself a distributary of the Mississippi River);
it consists of a series of five major bayous (slow-moving rivers or streams)
branching off from Bayou Lafourche in roughly the shape of a human hand.
Bayou Lafourche, the "thumb" of this configuration, is located in Lafourche
Parish. The remaining four major bayous—Terrebonne, Petit Caillou, Grand
Caillou, and Dularge—are situated in Terrebonne Parish, with smaller offshoot bayous like Pointe au Chien and Bayou Jean Charles forming bases of
settlement as well. Bayous Lafourche and Grand Caillou are wide waterways;
Bayou Pointe au Chien is about a third their width (and in many places not
much larger than a ditch), and Bayou Jean Charles is smaller yet, particularly
since its damming in 2007.

LAFOURCHE PARISH

Bayou Lafourche, nicknamed the World's Longest Main Street, is wide and
relatively densely populated[2] from its branch with the Mississippi at Donaldsonville (in Ascension Parish) to Golden Meadow, some one hundred
miles away. Some small areas of lesser density do exist: in Lafourche Parish,
they include a sparsely populated gap about ten miles in length between
Larose and Lockport through which the Intracoastal Canal (a portion of

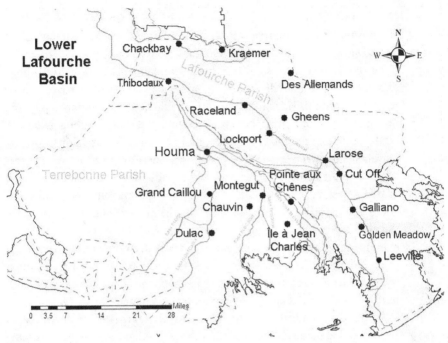

Figure 2.1. Map of the lower Lafourche Basin. Image created by the author.

the Intracoastal Waterway) passes just above Larose. The Intracoastal is thus an informal dividing line in Lafourche Parish, separating the Thibodaux/Raceland/Mathews/Lockport area ("up the bayou," i.e., upstream, upper Lafourche Parish) from "down the bayou"—the corridor comprising the towns of Larose, Cut Off, Galliano, and Golden Meadow, in order of descent (lower Lafourche Parish). Leeville, a tiny settlement with only a handful of residents, about ten miles below Golden Meadow, is the southernmost point of settlement on the bayou; Port Fourchon, another thirteen miles south, sits on the edge of the Gulf. Two roads flank the bayou: Louisiana Highway 1 on the west bank, and Louisiana Highway 308 on the east bank (also called East and West Main Street, respectively; in Golden Meadow, LA 1/West Main Street is called Bayou Drive). The roads follow the bayou nearly to its mouth—LA 308 ends at Golden Meadow, but LA 1 continues through to Grand Isle and the Gulf—and in general only small lanes branch off from the main road, typically leading back at most a few hundred feet to homes that often belong to family members of those on the bayouside. This pattern is characteristic of the region (and of French *habitant* settlement patterns in general in the

New World). Pointe aux Chênes and the Île à Jean Charles, however, have significantly fewer side roads; in fact, the Island is so narrow today that it has no side roads at all. In any event, directions in the Lafourche basin are not given using a north/south/east/west orientation. Direction is relational, and the four "cardinal" directions are up the bayou, down the bayou, across the bayou, and this side of the bayou, with magnetic orientation depending on which bayou one is on (though, generally, up the bayou is roughly north).

Bayou Lafourche today is a busy, modern thoroughfare, dotted by restaurants and especially fast food franchises (there are two McDonald's restaurants in the Larose–Golden Meadow corridor alone) and amenities such as banks, restaurants, shopping plazas, large supermarkets, car dealerships, and the local radio station (KLRZ 100.3). The placement of Port Fourchon, the base for most US oil production in the Gulf of Mexico, at the lower extreme of the bayou, has contributed to the waterway's importance. While the economy of the other three bayous is largely based in fishing (though many residents across the region work in the oil industry, regardless of which bayou they live on), Lafourche's economy is more broadly based.

In the past, South Lafourche was relatively segregated, with American Indians living below the main settlement at Golden Meadow at the southern end of major development and attending a separate school intended for Indian children only; the school was established around 1940 and ran to the seventh grade. The small white schoolhouse, known locally as the Settlement School, still stands and today functions as the headquarters for the United Houma Nation in the area. The residents living around the school are still primarily American Indian. The bulk of the rest of the population is white—and, indeed, the overall makeup of both South Lafourche and South Terrebonne is predominantly white— with only a smattering of black, Asian, and Hispanic residents intermixed. Especially since the 1960s, Indians have also moved up the bayou, especially to Cut Off, Galliano, and Larose.

Running parallel to LA 1 from Larose until below Golden Meadow where it merges with LA 1 is a four-lane highway with a large neutral ground (median), Highway 3235. Passing behind the bulk of settlement and thus primarily through nearly empty fields, it was so new when I first began visiting the area that it did not appear on my map, though it did sport a car dealership, a Walmart, a McDonald's, and the area's newest and shiniest motel, a Days Inn, among a handful of other businesses. In recent years, however, it has seen a minor explosion of development, with new supermarkets, gas stations, restaurants, and other businesses beginning to fill in the once-rural landscape. Within Golden Meadow town limits, the highway is called Alex

Plaisance Boulevard. All this is seemingly in defiance of the obvious signs that serious erosion is underway: there is often water on the road on LA 1 in Golden Meadow when a south wind is blowing at high tide, and there is no more ubiquitous sign of land subsidence than the chain of partially submerged telephone poles that runs alongside the road from Golden Meadow to Port Fourchon.

TERREBONNE PARISH

The development on Bayou Lafourche is somewhat in contrast to the situation in Terrebonne Parish. While Lafourche presents a long, nearly unbroken chain of both commercial and residential buildings, the lower bayous in Terrebonne Parish are almost exclusively residential, with few major commercial enterprises of any sort below Prospect Road on the south end of Houma, the Terrebonne Parish seat and the largest city in the area (population roughly thirty-four thousand). Though this is less true of Bayou Grand Caillou and Bayou Terrebonne around Chauvin, the area has a much more rural feel to it than does Bayou Lafourche. Residents in lower Pointe aux Chênes and the Île à Jean Charles live anywhere between seven and twelve miles from the nearest gas station, and the only restaurant serving the lowest reaches, at the Pointe aux Chênes marina, closed following Hurricanes Katrina and Rita in 2005 and never reopened. H's Corner, a gas station in upper Pointe aux Chênes that supported a small po-boy shop as well, closed following Hurricane Ike in 2008, though it has since reopened as CC's Grill. In recent years, a small sandwich shop has opened adjacent to the small grocery store at the lower end of upper Pointe aux Chênes, and small stores and a few restaurants are also available up the bayou in Bourg. A few restaurants and other services (for example, banks) are also found in Chauvin and Bourg. Larger chains can be found in Chauvin (Piggly Wiggly) and Houma (Rouse's), the latter roughly a half hour's drive away.

Pointe aux Chênes

The town of Pointe aux Chênes is located on the banks of Bayou Pointe au Chien, an offshoot of Bayou Terrebonne. A two-lane road, Highway 665, splits off of Highway 55 just above Montegut and follows bayou Pointe au Chien through the town of Pointe aux Chênes. The town itself is divided into two sections: upper Pointe aux Chênes and lower Pointe aux Chênes, the former beginning roughly at the split of the road and running about six miles down the bayou, and the latter then separated from the former by about two miles of open marshland. In lower Pointe-aux-Chênes, a small single-lane road, Oak Point Road, branches off from Highway 665, crosses over the bayou, and then follows on the opposite bank. The town is in fact split between two parishes: residents living on the left bank of the bayou are technically in Lafourche Parish (the parish line runs down the center of the bayou), while those on the right are in Terrebonne. In practical matters, the Lafourche Parish residents attend Terrebonne Parish schools and generally work and live in Terrebonne Parish. They vote in Lafourche Parish elections, however, and occasionally the matter of their parish affiliation comes into play when receiving—or not receiving—services. In the wake of Hurricanes Gustav and Ike in 2008, for example, I drove down to help out with the cleanup. When a friend from Golden Meadow and I stopped at a distribution center for emergency supplies in Galliano, we were asked to prove we were residents in order to receive provisions. After we did so, we announced that we were taking the goods to Pointe aux Chênes, causing the official in charge to blanch and widen his eyes: "The Lafourche side of Pointe aux Chênes . . . ," he said with an air of horror, clearly just realizing that it had been forgotten. He helped us load up some boxes and called out for updates on the town as we left.

In the past, this distinction had more frequent impact: schools in Lafourche Parish allowed American Indian students to integrate white schools before those in Terrebonne did. Consequently, some Indian teenagers living on the Lafourche side of Pointe aux Chênes in the early 1960s attended high school in Larose, though the trip was so long and the experience so unhappy that only one of the group of four I interviewed graduated. Like Golden Meadow, Pointe aux Chênes was relatively segregated, a situation that is better reflected in the modern population distribution than it is on Bayou Lafourche: lower Pointe aux Chênes is populated almost exclusively by American Indians. In the past, upper Pointe aux Chênes was occupied nearly exclusively by Cajuns. In recent years, however, many Indians have moved into upper Pointe aux Chênes and beyond to Bourg or even all the way into Houma (and some have gone well beyond, of course), for various reasons including but not limited to response to coastal erosion and the destruction of their homes, a desire to be close to their places of employment, and marriage into the white community. The two sections of the bayou are visually distinct as well: upper Pointe aux Chênes is populated primarily by single-level brick or wood houses with yard space, generally along the

Figure 2.2. Houses on stilts, Pointe aux Chênes. Photos by the author.

bayou, though a few small streets branch off the main road. In lower Pointe aux Chênes, there are no side streets (the narrow strip of available land that remains does not permit much depth of development), the houses are uniformly made of wood, and most strikingly, most homes are raised ten feet or more on stilts (see figure 2.2). The elevation of homes is a relatively recent development, a response to hurricane flooding in the area in the past twenty to twenty-five years. In fact, the Indian population formerly lived below the Cut-Off Canal (the canal that today runs along the lower end of lower Pointe aux Chênes), and the modern site represents a move in the mid-twentieth century away from land rendered uninhabitable by encroaching saltwater. In any case, as time goes on, the elevating of houses is becoming less of a distinguishing factor between the lower and upper sections of the town as increasing flooding motivates residents higher and higher on the bayou to raise their homes. Much like the submerged telephone poles along Bayou Lafourche, the raised homes are a constant reminder of the encroaching Gulf, sometimes rendered almost comical: when trailers are elevated, they are almost always encased by large beams to prevent them from blowing off in a strong wind.

Île à Jean Charles

Like lower Pointe aux Chênes, the settlement at the Île à Jean Charles represents a move to higher ground from an earlier settlement located further toward the Gulf. Today, the Island (as it is more commonly known) is connected to lower Pointe aux Chênes by a road about two miles in length connecting the two roughly a mile or two from the lowest reach of the town of Pointe aux Chênes. Also like in lower Pointe aux Chênes, the houses here are elevated, and settlement follows a very narrow strip along the bayou. With a population of only about seventy in 2017 (roughly a third of the number when I began fieldwork there), the Island is both the smallest and the lowest of the four towns; it falls outside the current levee protection area for Terrebonne Parish and is therefore susceptible to the most flooding and storm damage. Once several miles wide and home to herds of cattle, today the island is a sliver only a few hundred feet wide. Even early in my fieldwork, the road to the Island frequently flooded at high tide, and the road was so badly damaged by Hurricanes Gustav and Ike in 2008 that in places it was reduced to a single lane, the rest having washed away with the storm surge. (It has since been repaired.) In the summer of 2007, the bayou was dammed to prevent or at least limit flooding. In January 2016, however, the Louisiana Office of Community Development–Disaster Recovery Unit, working in partnership with the Biloxi-Chitimacha Confederation of Muskogees and the Lowland Center in Terrebonne Parish (Concordia et al. 2016:4), received a $48.3 million grant from the US Department of Housing and Urban Development through the National Disaster Resilience Competition to relocate the entire community to a new location. The members of the confederation have been heralded in the press as the America's first climate change refugees (e.g., Davenport and Robertson 2016; Van Houten 2016; Lagamayo 2017).

Grand Caillou/Dulac

Grand Caillou and Dulac are technically two towns located along the banks of Bayou Grand Caillou below Houma. In practice, however, it is hard to distinguish the two, as settlement is fairly dense for the length of the bayou and the only hint that one may have left one and entered the other is that the name on the water tower is different. Bayou Grand Caillou, on which they are situated, is a wide bayou, comparable in width to Bayou Lafourche. As on the other bayous, the Indian population is largely located on the lower reaches. But unlike Pointe aux Chênes and the Ile à Jean Charles, there is no clear separator (open marsh or a road presenting an easy line of demarcation) between the different ethnic populations. Consequently, the population is more integrated than those of the other Terrebonne communities. Grand Caillou/Dulac also reports a higher rate of Indian residents than does Bayou Lafourche: in Golden Meadow, only 4.8 percent of the population is Indian; in Dulac, this number is 39.4 percent (US Census Bureau 2000).[3]

SETTLEMENT HISTORY OF TERREBONNE-LAFOURCHE

Early History and Indigenous Settlement

At the time of European contact, Terrebonne and Lafourche Parishes were occupied by bands of Washa, Chawasha, and Chitimacha (Foret 1996; Kniffen, Gregory, and Stokes 1987). These populations were ultimately forced to relocate several times as a result of circumstances related to the European colonization of Louisiana. Some would become the target of European genocidal tactics—notably, the Chawasha were living on the left bank of the Mississippi River near New Orleans around 1860 when they were nearly annihilated by enslaved people who had been sent to wipe them out (Kniffen, Gregory, and Stokes 1987:79). However, the geography of the lower bayous provided an ideal hideout for remnant groups of American Indians escaping removal or extermination because of the lack of arable land potentially attractive to white settlers and the difficulty in navigating the lower bayous, a situation that persisted until the paving of roads in the twentieth century (Miller 2004; Fischer 1970; Stanton 1971, 1979). Small groups of Indians appear to have taken refuge in these bayous, and these groups merged over time to form one or more amalgamated cultural entities. Complicating the story of Indian settlement of Terrebonne-Lafourche, however, is the story of the movement of another Indian group, the Houma, in the late seventeenth and early eighteenth centuries.

The Houma, like many southern American Indians, had also relocated several times since their first contact with the French in 1682 (Bowman and Curry-Roper 1982; Parenton and Pellegrin 1950; Speck 1943; Stanton 1971; Swanton 1911). Originally settled on the banks of the Mississippi in either what is today northern Feliciana Parish or the southern edge of the state of Mississippi (documentary evidence indicates the former; however, archaeological evidence supports the latter [Bowman and Curry-Roper 1982]), decimation through contact with European disease and war with neighboring groups forced the Houma to move, first to Bayou St. John, today a part of the city of New Orleans, then to Ascension Parish to the west, and ultimately to an area near the modern site of Donaldsonville (Parenton and Pellegrin 1950; Rottet 2001; Stanton 1971, 1979). Throughout this turbulent period, the Houma integrated members of many southeastern coastal groups into their midst, including members of the Taensa, Bayougoula, Acolapissa, and Quinipissa, though the nature of the group is said to have always been understood to be predominantly Houma (Rottet 2001; Stanton 1979). There are two versions of what happened following their arrival in the Lafourche Basin.[4] The first version, the one generally accepted in the literature, is that as the Houma expanded south into their new territory, they met and ultimately added to their ranks the people already living in the lower bayous, especially the Biloxi, Chitimacha, Washa, and Chawasha (Rottet 2001; Stanton 1979). Some residents, however, contend that they did not and do not identify as Houma and that while Houma may be part of the ancestral mix, Houma identity was never dominant over that of the pockets of other refugees, among them mainly the Chitimacha but also the Biloxi, Choctaw, Acolapissa, and other groups (Pointe au Chien Indian Tribe 2005).

The earliest documented record of the ancestors of the modern American Indian groups in the lower Lafourche Basin comes from the early nineteenth century. Originally settled on Bayou Terrebonne, the Indian population began to relocate following the influx of Anglo-American settlers between 1810 and 1820. By 1836, Indians were living on land they had purchased from the US government on Bayou Petit Caillou, particularly below what is now the town of Robinson Canal, below Chauvin. The government had previously considered this land "uninhabitable" swamp (Westerman n.d.:19); when it was opened up for private purchase in 1820, Indians began to purchase the land on which they had been living for some time.

European Arrival

Around the time the Houma were making their way toward the Lafourche Basin, European settlers, primarily francophones, were also entering the area. Acadians, expelled from what is now Nova Scotia in 1755 and having spent a generation in France, arrived in 1785 and rapidly spread from their original settlements between Labadieville and Lafourche Crossing (roughly where the city of Thibodaux is now located [Oubre 1985]). They were joined in their new home by earlier arrivals to Louisiana from the Acadian and German coasts (areas about thirty miles upriver from New Orleans, settled by members of those groups) and, between 1810 and 1820 in the wake of Louisiana's statehood, by a surge of anglophone Americans, especially from Concordia Parish, Natchez, and New England (Oubre 1985). Small numbers of Portuguese, English, and Spanish immigrants further swelled the ranks. By 1842, Golden Meadow had been established. These people were also joined by large numbers of French settlers of non-Acadian origin; in Lafourche Parish, this group included refugees from the towns of Chenière Caminada and Leeville, which were devastated by hurricanes in 1893 and 1915, respectively. (Leeville was founded by survivors of the hurricane that devastated Chenière Caminada.) After those hurricanes, people all but abandoned those communities and moved up the bayou, settling largely in the corridor between Golden Meadow and Larose (Pitre 1996). French became the dominant language in the area (until it was replaced by English in the twentieth century), including among the American Indian population. Acadians thus were not the only francophone group to enter the region, nor were they isolated there, as the literature often contends (see e.g., Caldas 2007). In fact, the immigrants of French origin who entered the Lafourche basin in the nineteenth century ultimately outnumbered the Acadian settlers. In Lafourche Parish, those of French origin came to outnumber settlers of Acadian origin at a ratio of 3:2, while in Terrebonne, the ratio was 2:1 (Brasseaux 1992:167). Family names in the Lafourche today reflect this history: approximately 45 percent of names listed for the area in West's (1986) *Atlas of Surnames of French and Spanish Origin* are Acadian (see chapter 3). The original Lafourche Parish was divided into two, creating Terrebonne Parish, in 1832. Today both parishes are among the most francophone in the state (US Census Bureau 2000).

The Meeting of Old and New World Populations in the Lafourche Basin: From Plantations to Federal Recognition

Once land was made available for sale on the lower reaches of Bayous Terrebonne and Petit Caillou in the early nineteenth century, both indigenous and European settlers quickly snapped it up, and by 1846, the Indians on lower Petit Caillou had been joined by both small-scale white farmers and planters in search of land for sugar production. That year, James Baker Robinson, a planter who arrived from Mississippi with his wife and children (US Census Bureau 1850a), bought land five plots above those purchased ten years earlier by ancestors of the modern Indian communities and just below the canal that bears his name today (Westerman n.d.). By 1849, Robinson was operating a plantation producing two hundred hogsheads of sugar per year on land worked by the nearly sixty enslaved people he employed (Champomier 1846–62; US Census Bureau 1850b). Robinson's story is a fairly common one. In fact, sugar became a major product of the lower Lafourche Basin, with plantations or small-scale individual family producers on all four major bayous of Terrebonne Parish and on Bayou Lafourche as far south as Larose (Champomier 1846–62). Indians cashed in on the new crop as well; in 1851, Celestin and Alexandre Billiot, ancestors of the modern Indian community, opened a small sugar mill on Bayou Pointe au Chien. While their farm never produced more than sixty hogsheads annually in the decade it was in operation (and that number is exceptional; most years they produced between fifteen and twenty hogsheads [Champomier 1846–62]), major plantations did arise on the larger bayous. Bayou Grand Caillou had many large plantations, each employing hundreds of enslaved people and producing thousands of hogsheads of sugar annually (Champomier 1846–62; L. Bouchereau 1869–78, A. Bouchereau 1878–1918; Menn 1964). While a smaller plantation, Hard Scrabble Plantation, was established at the fork of Bayous Terrebonne and Pointe au Chien in the post–Civil War period (A. Bouchereau 1878–1918), the sugar industry in Terrebonne Parish generally was limited to the four major bayous and within that region primarily to Bayou Grand Caillou. J. B. Robinson's plantation at Robinson Canal was only a major producer for the first ten years of its operation; by 1859, Robinson had moved the bulk of his operation to new land at what would become Cedar Grove Plantation on Bayou Grand Caillou, and sugar production on Bayou Petit Caillou moved to plantations further upstream (Champomier 1846–62; Menn 1964).

Indigenous Movement in the Nineteenth and Twentieth Centuries

At the same time that Robinson was establishing and moving his sugar business to Bayou Grand Caillou, his Indian neighbors were beginning their own migration. By the early 1850s, a few families had migrated from the Bayou Petit Caillou/Bayou Terrebonne region and had established themselves at Pointe aux Chênes. Land was made available for sale on the Ile à Jean Charles in the 1880s, and the early purchasers were almost uniformly Indians, offshoots of the Terrebonne/Petit Caillou group that had first moved to Pointe aux Chênes. A large contingent of the original group also went in the other direction, to Bayou Grand Caillou, where they lived first at Four Point Bayou and Bayou Salé on the lower reaches of the bayou before moving further up the bayou. The original settlement below Robinson Canal had been almost entirely abandoned by Indians by 1910. In their new homes, the Indians supported themselves primarily through fishing and trapping; especially on Bayou Grand Caillou, they also supplemented their income by working the agricultural fields, including cane-producing fields, for wages (Westerman n.d.).[5]

As early as 1850, a few Indian families had also made their way toward Bayou Lafourche, settling in the swamps between Bayous Lafourche and Pointe au Chien (Westerman n.d.). Hurricanes, especially that of 1909, sent them further inland toward Golden Meadow, where they settled in the area below town today known as the Indian Settlement. The bulk of the Indian population in Golden Meadow, however, came in the early twentieth century, when large numbers of other refugees from Pointe aux Chênes and the Île à Jean Charles, again escaping hurricane-damaged homes, made their way there. Until the early twentieth century, Bayous Pointe au Chien and Jean Charles were occupied several miles below the modern settlement areas, below what is now the Cut-Off Canal. Residents moved up the bayou or toward Bayou Lafourche in response to natural disasters. In the 1960s, Indians began migrating from the Golden Meadow area up the bayou as far as Larose, with a few relocating as far away as Houma and Thibodaux. Today, the Indian population is fairly evenly distributed along the Larose–Golden Meadow corridor (US Census Bureau 2000). A rivalry developed between the different Indian communities, and though intermarriage was (and is) common, so were teasing and fighting. The residents of lower Pointe aux Chênes were called by their neighbors the *Pointus*, the Islanders *Buffalos*, those from Dulac/Grand Caillou were the *Mirlitons*, and those from Bayou Lafourche the *Fourchus* or *Fourchons*. When I asked about the nature of these rivalries, I was always given these nicknames with a great laugh and assurance that the rivalries were largely friendly though at times somewhat violent. (I heard stories of what could perhaps best be described as rumbles.) On the other hand, rivalries between ethnic groups on the bayou were not generally so friendly. When I asked what people of upper Pointe aux Chênes were called, I was told either that they were Cajuns or that they were *Blue Bellies*, a pejorative term. Quite possibly, that may have been a response to pejorative terms aimed at Indians by the Cajun population. Cajun hostility toward Indians was previously very high, and while it has not disappeared, the situation today is clearly much less strained than it was fifty years ago.

Indigenous Issues in the Twentieth and Twenty-First Centuries: Struggles over Identity

The Indians of the Lafourche Basin began their now well-documented struggle for recognition from the federal government in the 1970s. In 1972, Helen Gindrat founded Houma Indians Inc. in Golden Meadow. In 1974, a parallel organization, Houma Alliance Inc., was founded. The two merged to become the United Houma Nation (UHN) in 1979 and that year filed a letter of intent to petition for federal recognition with the US Bureau of Indian Affairs; the petition was filed in 1985 (United Houma Nation n.d.). In 1994, the UHN's petition for federal acknowledgment received a negative proposed finding on the grounds that three of the seven criteria necessary for recognition had not been met; most notably, the bureau felt that the group could not trace itself to a single historic Indian group. However, the finding also suggested that the UHN in fact comprised several separate independent communities (US Bureau of Indian Affairs 1994). Meanwhile, in 1993, the Pointe au Chien Indian Tribe, in recognition of its claim to primary descent from the remnant populations of diverse origin taking refuge in the lower bayous, adopted a constitution and filed articles of incorporation as an autonomous Indian community with the Louisiana secretary of state, thereby formally establishing itself as a separate entity from the UHN (Ferguson n.d.). The Biloxi-Chitimacha Confederation of Muskogees, comprising three bands, one in each of Golden Meadow, the Île à Jean Charles, and Dulac/Grand Caillou, filed its own articles in 1995 and consolidated into a confederation and submitted a letter of intent to the Bureau of Indian Affairs that same year (Fleming 2013). The Pointe au Chien Indian Tribe filed its letter of intent in 1996 (Fleming 2013). All four petitions received negative preliminary findings in May 2008 though the findings differed somewhat from those of the UHN.[6] As of this writing, all of these findings are pending appeal, but proceedings have been suspended indefinitely by the imposition of a state of emergency following the 2008 hurricane season.

The Aftermath of Plantations: Contact between the Free and the Enslaved and Its Consequences

On the eve of the Civil War, enslaved people comprised 56 percent of the total population in Terrebonne Parish and 46 percent in Lafourche (US Census Bureau 1860b). Their descendants are important in this context for two reasons: first, to give a complete picture of the demographic and linguistic history and modern context of the area, and second, to include the possibility that Louisiana Creole (see chapter 3) might have been spoken in the lower bayous (given its association with slavery), which would present the potential for influence on the extant varieties of French. (Creole is not currently spoken in lower Terrebonne-Lafourche.) While nearly the entire modern population of the lower bayous is either Cajun or Indian, the area historically had a large black population, which may have affected the modern linguistic situation.

As a major sugar-producing region, the Lafourche basin benefited from the labor of thousands of enslaved people. Most if not all the plantations in the lower Lafourche were established after 1820, and while the industry suffered during and after the Civil War, causing the closure of several plantations (Heitmann 1996:99), it rebounded. Sugar continues to be produced today, although improvements in production technology (American Sugar Cane League 2018) mean that only two sugar mills remain in the lower Lafourche basin (in Raceland and Thibodaux). It seems a bit of an anomaly that Terrebonne-Lafourche currently has very few black people living along the lower bayous and only a handful of others in the area at large (African Americans now constitute 20 percent of Terrebonne's population and 14 percent of Lafourche's [US Census Bureau 2016]). The modern black community is concentrated in the urban areas located in the upper reaches of the region. Houma, the largest city in the area and the Terrebonne Parish seat, has a sizable black population at 24.4 percent (US Census Bureau 2010a).[7] Thibodaux, the only city in Lafourche Parish (though half the size of Houma, with a population of 14,566) and the parish seat, has that parish's highest concentration of black residents (32.8 percent), with neighboring Raceland close behind (28.5 percent) (US Census Bureau 2010a). The highest concentration in the area is in the town of Gray, located in the corridor between Houma and Thibodaux, where the rate reaches 37.9 percent (US Census Bureau 2010a). Still, given the historical presence of black people, these numbers are strikingly low, and beyond the city limits, the concentrations of black residents rises above 2 percent in only two places: Larose (5 percent) and Lockport (2.4 percent) (US Census Bureau 2010a).

A quick examination of the historical US census records (Pierron 1942, table 15) explains to some degree what happened to create the situation. Following the Civil War, a mass exodus of formerly enslaved people from the lower bayou and in fact from the region as a whole occurred, resulting in a sharp drop in the black population between the 1860 and 1870 censuses. This exodus likely resulted at least in part from the sugar industry's relocation to the Caribbean, which resulted in the shuttering of plantations in the Lafourche country (Heitmann 1996:99). Two further drops in the population occurred in the 1910s and 1920s; the latter may also have been precipitated by the collapse of the sugar industry in 1920–21 (Heitmann 1996:101). That said, other factors clearly contributed to the exodus. Racial tensions ran high in the Reconstruction era and resulted in lynchings and events such as the Thibodaux Massacre, an 1887 sharecroppers' strike that resulted in the deaths of dozens of black residents. Moreover, local lore suggests that following the war, bitter white residents of the lower bayous were eager to drive out the newly free blacks and consequently refused to hire them. Quite possibly, then, the exodus was just as much forced as voluntary. Those who stayed in the area appear to have generally migrated toward Houma and the Thibodaux-Raceland area, and today, asking where one might find black people to interview results in the response, "Go to Houma!"

That said, the presence of a sizable population of enslaved people presents the potential for the presence of Louisiana Creole and for its impact on the French spoken by other groups in the lower Lafourche Basin. Determining whether this is a plausible scenario here is complex: first, the presence of Creole must be established, and then evidence of significant interaction between free people and the enslaved must also be present. The historical record suggests that the latter is quite possible: in Lafourche Parish, 30 percent of enslaved people lived on family farms or in towns, often sharing dwellings with slaveholders, rather than on plantations (Michot 1996:31–33; US Census 1860a). In Terrebonne, only 12 percent of the enslaved lived with their captors; however, 32 percent of the free population in Terrebonne employed slaves. Equally important, while large plantations employing hundreds of enslaved people did exist, the vast majority of slaveholders in Terrebonne-Lafourche were individuals or families that held between one and nineteen people (Michot 1996:34), with more than half holding fewer than six slaves (Michot 1996:32). With one-third of the population holding slaves and only a small fraction of that number running large-scale plantations, close interaction between people in bondage and their captors or free servants (for example, Irish immigrant women who were hired as house servants for slaveholders [Michot 1996:37]) is indicated and would lay the groundwork for several linguistic scenarios, including the transmission of French to enslaved anglophones and transmission of Creole from enslaved creolophones to their francophone or anglophone masters.

It seems unlikely, however, that Creole had much of a presence. The modern black community appears to have been monolingual in English for some time. Despite my constant efforts, my research found only one black francophone resident—an elderly man who had learned French as a second language from his Indian and Cajun coworkers as an adult. The difficulty in finding francophone black people points to a history in which English, rather than a variety of French, predominated for that population. Further examination of the historical record suggests that if African Americans did speak French, they usually acquired it from the surrounding white population. Evidence comes from several sources, none of which is conclusive on its own, but in combination, they provide strong corroborating evidence. These include bills of sale detailing the origins of enslaved people and the names of the slaveholders involved, fugitive slave ads, and given names, which also appear in such sources as church baptismal, marriage, and death records. Using these multiple lines of evidence, Picone (2003) makes a strong case for the presence of a large anglophone contingent of enslaved people on the Cane River and suggests that the same was true elsewhere. Klingler (2003a:107–8) confirms the presence of a large number of anglophone slaves in the False River area in Pointe Coupee Parish. The data I have collected on the Lafourche Basin support this suggestion for Terrebonne-Lafourche as well. The elite planter class in Terrebonne-Lafourche consisted of anglophones (Michot 1996:37) and upwardly mobile Acadians who rapidly assimilated to the Anglo-American norm (Brasseaux 1996). These people seem to have brought a good number of enslaved people with them from their places of origin or purchased them later, whether in New Orleans or directly from the eastern states after the importation of enslaved people from Africa and the Caribbean became illegal in 1808. Moreover, while some of the existing enslaved population clearly was native to Louisiana, supply could not keep up with demand: Robert Ruffin Barrow, who owned several large plantations in Louisiana, including the massive Caillou Grove Plantation on Bayou Grand Caillou, made "an extended slave-buying trip to Virginia" in the summer of 1834 with his brother, William, in search of enslaved people to bring back to Louisiana (Barrow Plantation Journal; quotation from archivist's description). Likewise, just over fifty of the people included in the sale that kept Georgetown University (then Georgetown College) afloat in 1838 were sent to Terrebonne Parish (DeSantis 2016; Duchmann 2017; Parker 2017). Entries in the plantation journal kept by Barrow's overseers at Residence Plantation in Houma document the acquisition of more likely anglophone slaves in 1857 (Barrow Plantation Journal). A July 10 entry records the arrival of thirty-four enslaved people purchased in New Orleans, all of whom bear English names. An English name is hardly irrefutable evidence of English-speaking ability, of course, but given that the enslaved by and large named their own children (Inscoe 2006), it is reasonable to assume that people named Nancy, Philis, Martha, Phoebe, and Jim were anglophones. Perhaps the most suggestive evidence for the anglophone nature of the enslaved population comes from a fugitive slave advertisement placed by Robert H. Barrow (possibly a typo) in 1854. Fugitive slave ads are a particularly useful window into the linguistic practices of the time. The ads were placed either by slaveholders looking to retrieve fugitive slaves or by people seeking those who claimed ownership of captured runaways. Those taking out the ads generally described fugitives' appearance, clothing, and sometimes even demeanor, and some of the descriptions are quite detailed, occasionally including the language spoken by the fugitive. The two people Barrow was seeking spoke "very indistinctly, [their] pronunciation being that of a Virginia or Carolina negro" (R. H. Barrow 1854). In other words, their speech was like that of most other enslaved Louisianans—English.

The predominantly anglophone nature of the enslaved population is borne out by the 1870 census, the first following emancipation. It documents the first names of all residents of the parish and includes the formerly enslaved.[8] A search of names of those listed on the census as black reveals predominantly English names: of the first one thousand people listed as black on the census in Terrebonne Parish, only sixteen are unambiguously French (e.g., Toussan Alvatore, Ursin Areille, Batiste Broux), and some have English last names (implying that they or their forebears had been in the employ of anglophone slaveholders), while others have French surnames (implying that they were formerly held by French slaveholders). Even those with French names were unlikely to give their children French names, however, instead choosing clearly English names such as James, Ellen, and Mollie. Six of the sixteen are children under the age of eighteen; none of them have siblings with French names. This evidence points to predominant use of English by the community.

Evidence does support the presence of at least a few francophones among the enslaved as well, however. Some of these people were likely native francophones. Again, a simple overview of first names serves as a starting point. A number of enslaved people documented in church baptismal, marriage, and death records for the Lafourche basin have unambiguously French names like Angelique, Agathe, Eulalie, Hypolite, François, Jean Pierre and Louis Monet, which, again, while not irrefutable evidence of linguistic ability suggest that they—or at least their parents—may have spoken French.[9] Further support comes from conveyance records—documents recording legal transactions including the exchange of enslaved people. In Terrebonne Parish, for example, conveyance records from 1824 through 1826 (Terrebonne Genealogical Society 2019) document several instances of transfers of slaves, often with French names themselves, from one French-named slaveholder to another. For example, Act 202 shows that on April 21, 1824, Alexandre Dupré mortgaged $450 to Henry S. Thibodaux to pay for the purchase of a twenty-year-old woman, Pouponne, from someone named Renaud *fils*. And on May 19 of the same year (Act 210), Joachin Porche transferred to H. S. Thibodaux two people, Magloire and Ursule. Several of the enslaved who appear in these documents were very young (including one child aged only ten), and almost all of the slaveholders have French names as well, implying their own use of that language.

That said, it seems far more plausible that if the enslaved spoke French, they did so because they were anglophones who picked up French as a second language from the free population when they arrived in Terrebonne-Lafourche. French was certainly the dominant language in the area: the Barrow Papers at Tulane University contain a notebook belonging to Robert Barrow Ratliff, a relative—he appears to be a nephew—of Robert Ruffin's, for example, showing that someone in the family studied French and may well have had a good command of the language). Furthermore, Heitmann (1996) notes that the elite planter class to which Barrow belonged represented less than 1 percent of the slaveholding population and that while the elites were anglophone, they stood in contrast to their predominantly French Catholic neighbors—many of whom were also slaveholders, though on a smaller scale. A review of the 1860 slave schedule (US Census Bureau 1860a) reveals slaveholders with unambiguously francophone names such as Auguste Babin, Seraphin Gravois, and Narcisse Boudran (possibly a misspelling of Boudreau) alongside likely products of intercultural marriage such as Theodule Stevens and John Berger. These small-scale slaveholding contexts, in which the enslaved often lived in the same homes as those who claimed ownership of them, in towns rather than in agricultural contexts, are sites where the acquisition of French may well have occurred.[10] Evidence exists to support the contention that at least sometimes it did.

On January 4, 1859, a man with the very English name Williams was apprehended in St. John the Baptist Parish. The ad placed to alert the public of his arrest states that he claimed to belong to a Mr. Lorett of Houma (Jacob 1859). A review of the 1860 slave schedule suggests that Mr. Lorett was probably either Alex or Nicolas Lirette, both almost certainly francophone, who held twenty and two slaves, respectively. Williams is described as speaking both English and French. Likewise, in 1855 a suspected fugitive bearing the English name of Fleming is described as speaking both languages (Trépagnier 1855). An 1852 ad in French placed by Jean Rousseau in search of three fugitives named Jarry, Potney, and Bob noted that they speak "anglais et un peu le français [English and a little French]." Given the fugitives' (lack of) fluency in French, it stands to reason that they had learned the language via the francophone Rousseau and/or his associates.

In short, it seems very unlikely that a large creolophone population existed in lower Terrebonne-Lafourche. Today, as in the past, Creole is spoken in the town of Kraemer, more commonly known as Bayou Boeuf (due to its location there rather than on Bayou Lafourche), in upper Lafourche Parish; it is also spoken twelve miles north of Kraemer in Vacherie (St. James Parish), and twelve miles east in Des Allemands (on the St. Charles–Lafourche boundary). Interviews with residents of Kraemer and Des Allemands suggest that the three creolophone communities were historically linked, culturally and economically. That Kraemer is creolophone is well known to residents of the surrounding area, who immediately identify it when asked where people might speak differently. The town does not seem to be on the radar for those in lower Lafourche, however, who instead cite Grand Isle to the South (or neighboring lower bayou towns, for that matter) when asked the same question. The road leading to Kraemer is small and winds through densely forested areas; it is not unreasonable to assume that the lack of mention the town receives in lower Lafourche and Terrebonne Parishes results from its relative isolation from the lower bayou, both currently and historically.

While it is clear that at least some of the enslaved in lower Terrebonne-Lafourche spoke French, what they spoke, by and large, was more likely an L2 variety picked up from members of the surrounding francophone population, whether the slaveholders with whom they shared homes or other community members. Some may well have picked the language up after emancipation. Interviews I conducted in Thibodaux with people who grew up in mixed-race neighborhoods suggest the previous existence of a black francophone population in the area; my interviewees said the black residents spoke the same way they did. Given that people readily identify creolophone Kraemer as a place where people speak differently, it seems fair to take such assertions at face value. In any case, the fact that I could not find a single black native speaker of French implies strongly that the documentary evidence suggesting the predominance of English is accurate. While some African Americans clearly learned French, most preferred to speak English, and their linguistic role, as proposed by Picone (2003), was not to provide influence from Louisiana Creole but rather to provide pressure from below for a community-wide shift to English. Indeed, they may be the source of many of the anglicisms found in the French of the area. Any similarities between Louisiana Creole found in the French spoken by the current Indian or Cajun residents of the area must therefore be attributed to (1) commonality between the dialect and the one that lexified Louisiana Creole, (2) features that are typical in situations of dialect contact, (3) similarities between the processes of creolization and language shift, or (4) influence from Louisiana Creole at some time prior to their arrival in the area—for example, the Houma spent time on Bayou St. John near New Orleans (today part of New Orleans) and in St. James Parish prior to arriving in the Lafourche country. They may well have come into contact with Creole at either of those locations, given their proximity to enslaved Africans in the urban context (New Orleans) and later as a consequence of living on the borders of plantations and frequently entering the plantations to perform agricultural work or to hunt for pay (Kniffen, Gregory, and Stokes 1987:93; Dardar 2005:9). Indeed, given that some of the features that distinguish French as it is spoken by Indians (see chapter 4) are shared with Creole as it is spoken in Kraemer, which shares a historical connection to St. James Parish, this scenario is quite possible.

SUMMARY

The demographic history of the lower Lafourche Basin shows a population dependent on the land for its livelihood but also in constant flux, with a long history of moving from place to place in response to changes to the land or to the industries that rely on it. Ironically, some of the economic activities on which the people rely have also been at the source of the erosion of the land—most notably, the cutting of canals for the oil industry (see chapter 1). The interactions between groups—white, black, and indigenous—set the stage for the modern ethnic divisions in the region and laid the groundwork for the nature of the language spoken in the area, which would come to be French for all groups who remained in the lower bayous following the Civil War. Starting in the nineteenth century but increasing in speed in the twentieth, that language would find itself as endangered as the land on which its speakers live.

Chapter 3

THE HISTORY OF FRENCH IN LOUISIANA

The linguistic foundation of place in Terrebonne-Lafourche is the French language, which retained a hegemonic presence well into the twentieth century. Understanding the role it plays in place-making requires understanding its development in Louisiana at large and in the Lafourche Basin more specifically. Outlining its development can also dispel a few myths regarding French in Louisiana that persist despite both scholars' and laypeople's best efforts to dismiss them. Most important, the variety of French that is at the heart of this study, the one spoken in lower Terrebonne-Lafourche, is most definitely French. While it has been influenced by its intensive contact with English and to a lesser degree by its contact with other languages, it remains fundamentally French on every level.[1]

OVERVIEW: VARIETIES OF FRENCH SPOKEN IN LOUISIANA

The story of French in Louisiana begins with French colonization in the early eighteenth century. Despite the colony's transfer to Spain only sixty-three years after its inception and sale to the United States in 1803, significant francophone immigration continued until the eve of the Civil War. During these 160 years, francophones from many regions and walks of life made their way to the colony and, along with the non-francophone groups who joined them (whether by will or by force), shaped the rich and diverse linguistic landscape of the state.

Linguists have traditionally recognized three varieties of French in Louisiana: Colonial French, Louisiana Creole, and Acadian or Cajun French. The term *Colonial French* most accurately refers to the French spoken by the earliest settlers to arrive in Louisiana (i.e., during the colonial era); however, the term has also been used to refer to the variety that was brought to the state in the postcolonial (postpurchase) period. Nearly identical to modern Standard French save a few lexical items (for example, *banquette* [sidewalk];

cf. *trottoir*), the latter was brought to Louisiana by middle-class immigrants fleeing revolutionary and postrevolutionary France, planter-class immigrants from Saint-Domingue (modern Haiti), and by the children of the local planter class, sent to study in France, who carried the standard dialect they learned in school back to Louisiana with them. Given that the French spoken by the earliest colonists was linguistically distinct from the standard-like newer import, to alleviate confusion, a new term was introduced to designate the later-developed variety: *Plantation Society French* (Picone 1998, 2015). Colonial French, then, was a nonstandard variety that served as the base for the development of the two remaining varieties of French. Ironically, though Plantation Society French is the most prestigious and the most recently arrived, it also the most nearly extinct. This variety was used in newspapers and is the basis of a rich literary tradition in Louisiana; today, however, at best a handful of speakers and/or semispeakers remains.

The next variety, likely spoken by fewer than seven thousand people today (Klingler 2019), is Louisiana Creole. Though it is similar to Haitian Creole, historic and linguistic evidence (see, e.g., M. Marshall 1989; Klingler 2003a; Klingler and Dajko 2006) suggests that it is indigenous to Louisiana, the creation of the enslaved peoples who arrived from Africa and were obligated to learn the language of those in power. At the time of their arrival during the French and Spanish colonial periods, that language was (Colonial) French.[2] Lexically, it is nearly identical to other varieties, the primary exceptions being the verbs *gen* (to have) (cf. *avoir*) and *olé* (to want; cf. *vouloir*). It is only very slightly different on the phonetic/phonological plane, attesting, for example, higher rates of front vowel unrounding (/i/ for /y/, /e/ for /ø/, and /ɛ/ for /œ/, so that *du feu* [dyfø] becomes [dife], for example) and postvocalic *r*-lessness than other varieties (Klingler 2014). It is more readily identified by its morphological and syntactic attributes. Unlike other varieties of French, it is characterized by such features as verbs with no or limited inflection (some verbs have only one form regardless of tense and person, some have a long and a short form depending on tense); isolating morphology, including independent morphemes for number, tense, mood, and aspect; a lack of gender differentiation for both inanimate and animate referents (i.e., a single pronoun, *li*, for both male and female referents in the third-person singular); limited case marking in personal pronouns (the pronouns are all based in the objective form [see table 3.1 for a comparison of pronouns in the subjective case]); and reanalysis leading to the agglutination of nouns and their determiners (for example, *ma tête* [my head] becomes *mo latête*, essentially "my the-head").

Table 3.1 Subject Pronouns in Louisiana French and Louisiana Creole			
Louisiana French		**Louisiana Creole**	
je	on, nous-(autres)	mo	nous
tu	vous-autres	to	vous
il (m), alle/elle (f)	ils, ça, eux, eux-autres, eusse	li	yé

Given Louisiana Creole's strong similarity to other French varieties, the best way to identify a speaker of the language is to ask for a diagnostic sentence. In keeping with work conducted with Thomas A. Klingler, I ask speakers to translate "I have five dollars." In Louisiana Creole, this is rendered *mo gen cinq piastres*. *Cinq piastres* would be identical in the other varieties. (*Piastre* is an old French monetary term, borrowed from Italian, that is used across the state and in French Canada for *dollar*.) In the other varieties, however, the first-person subject pronoun is *je*, and the verb would be a conjugated form of *avoir*—in this case *ai*, producing *j'ai cinq piastres*.

While diagnostic sentences are generally useful, however, the line between Louisiana Creole and other varieties is not always clear. This is particularly true when speakers use features that may more typically belong to another variety (for example, I have documented instances of the Creole third-person plural pronoun *yé* appearing in speech that would otherwise clearly be classified as a non-Creole dialect), a phenomenon that is not uncommon in regions where people are bilingual in two varieties (such as Breaux Bridge, near Lafayette) and/or are creolophone but have contact with European and standard-like postcolonial Frenches (e.g., the town of Kraemer, in upper Lafourche Parish).

The third variety, spoken in the lower Lafourche Basin and the subject of this book, is the most widely spoken today; it is the variety most often called *Cajun French*. As was the case with Colonial French, however, this name can be misleading. It is known to linguists today as *Louisiana Regional French*, but speakers do not use this label. In Louisiana, people have a strong tendency to label their language in accordance with their ethnic identity. Consequently, someone who identifies as Creole most often calls their language *Creole* as well, even if linguistically they may speak what linguists would not classify as such. Likewise, people who identify as Cajun will call their language *Cajun French* even if structurally it would be characterized as *Louisiana Creole*.[3] This phenomenon is documented most extensively in Klingler's "Language Labels and Language Use among Cajuns and Creoles in Louisiana" (2003b). For this reason, when it is logical to do so, authors (e.g., Lindner 2019 conducting research on linguistic attitudes) continue to use the term in their research and writing. Until very recently, this was not an issue in lower Terrebonne-Lafourche (though it is certainly so in Kraemer, where white creolophones routinely call their language *Cajun French*); everyone I asked, regardless of ethnic affiliation, called their language *Cajun French* if they did not simply call it *French* (a term more common among Indian speakers). However, a speaker in an interview conducted by Klingler (and transcribed by me) in 2004 preferred to say he spoke *Indian French*; a few years ago, I began seeing sporadic references in print—most recently from the Pointe au Chien Indian Tribe on its Facebook page in 2017—to *Indian French* as well. Consequently, I follow Klingler (2003b, 2015) and other more recent authors in using the term *Louisiana Regional French*; when the term *Cajun French* appears, it indicates the language as it is spoken by those who identify as Cajun.

THE NATURE AND DEVELOPMENT OF
LOUISIANA REGIONAL FRENCH

Though it is popularly known that French is spoken in Louisiana and the term *Cajun French* is well known, even to outsiders, the nature and origins of this variety are poorly understood. Folk ideologies make many suggestions, sometimes contradictory. One oft-heard claim (usually the result of the pronouncements of a well-meaning French tourist with no linguistic background but some knowledge of older forms of the language) is that it is a relic of seventeenth-century French. Another claim is that Cajun French is so vastly different from the French spoken in France that the two are mutually incomprehensible. One also frequently hears that it is a mixed language, blending English and French elements equally (the same is said about Cajun English; see, e.g., Alvarez and Kolker [1988]; Cran and MacNeil [2005]). Most commonly, Louisiana Regional French is misunderstood to be a direct descendant only of the French of the Acadians who survived the Grand Dérangement in 1755 and made their way to Louisiana. Thus we find, for example, references to "Louisiana Acadian French" (Bodin 1987), or "The Acadian French of Lafayette, LA" (Doucet 1970). Scholars may sometimes contribute to the confusion by shorthanding Louisiana's complex history with such lines as "Cajuns are descendants of Acadians from the province of Nova Scotia in Canada who originally settled in Louisiana between 1765 and 1785" (Dubois and Horvath 2003a:192).[4] Still others suggest that Acadians remained isolated until the twentieth century and thus that their French is the only French to have survived in Louisiana until the present (Picone and Valdman 2005; Caldas 2007).[5] All of these suggestions are inaccurate, though they are based in truth.

Description and Origins

Louisiana Regional French differs in many aspects from modern Standard French. However, these differences are generally superficial. At the level of the sound inventory, they include such items as the pronunciation of *r* as an apical tap or flap (i.e., much like the central consonant in English *butter*) rather than a uvular approximant or trill (i.e., the Standard French pronunciation, though the latter is also attested in some parts of the state) and the retention of /h/ in certain words [e.g., *honte* [hɔ̃t] (shame) or *haut* [ho] (high)]). The front rounded vowels that give English speakers so much trouble are often unrounded in Louisiana. Thus, as is the case in Louisiana Creole, *tu* [ty] may be pronounced [ti] and *soeur* [sœɾ] becomes [sɛɾ]. Several phonological phenomena also serve to separate the two. Metathesis affects the words *je* [ʒə], *le* [lə], and *cette* [sɛt], producing [əʒ], [əl] (the vowel change due to unrounding), and [ɛst]; it also affects the prefix *re-*.[6] A number of morphological distinctions, amounting to the different conjugation of some verbs and the different gender classification of some nouns, like *ouragan* (hurricane) (feminine in Louisiana but masculine elsewhere), also serve to differentiate Louisiana French from that spoken in France. At the syntactic level, among other features we find the use of *être après faire quelque chose* (literally "to be after doing something") to indicate progressive action. In Standard French, progressive actions are indistinguishable from those that occur habitually, though it is possible to draw a distinction if necessary by using *être en train de faire quelque chose* for those that are actively ongoing without a break in action (i.e., progressive). In Louisiana, the habitual and the progressive have obligatorily different forms. Finally, a few lexical and semantic differences also exist. Thus, in Louisiana, one may be doing some-thing *asteur* (right now)(derived from *à cette heure* [at this time]), while in France one does it *maintenant*. In Louisiana, *espérer* means "to wait," but in France it means "to hope." Meanwhile, in France, *attendre* means "to wait," while in Louisiana it means "to hear." A French tourist asking where to buy *essence* will be directed to a pharmacy or department store in search of the perfume department instead of a gas station to fill up the tank on their *voiture*, which is a *char* in Louisiana.

In some cases, these differences are archaic or dialectal retentions, thus giving rise to the claim that the people of Louisiana speak seventeenth-century French. These include *être après faire quelque chose* and pronunciations such as the apical *r*. However, Louisiana French also possesses many features of modern Standard French (for example the pronunciation [wa] for orthographic *oi*, discussed below) that refute this claim. Other differences have arisen from the long-standing influence of English, thus feeding the assertion that the language is a mix of English and French. English is likely the source of such constructions as *ça faisait le bois durer plus longtemps* (it made the wood last longer; cf. *ça faisait durer le bois plus longtemps*) and is definitely the source of a significant number of lexical borrowings, some of which are ambiguously code-switches (see Picone 1994, 1997a), such as *truck* and *drive*, and some of which are fully integrated into French, such as *boulaille* (a bright headlamp) from English *bull eye*, and *coloille* (kerosene) from English *coal oil*. Though undeniably affected by contact with English, especially at the lexical level, Louisiana French remains fundamentally French, and it would be a stretch to suggest that it constitutes a mixed language any more than do other languages that have undergone sustained contact with and influence from another language. Qualifying as a mixed language would require systematic use of English forms in some domains and French in others. Instead, Louisiana French has an overwhelmingly French lexicon and structure with significant but nonetheless unsystematic English influence that represents a minority of forms.

Finally, a few differences result from borrowings from indigenous languages: in Louisiana, a *raton laveur* (raccoon) is a *chaoui*, from the Choctaw word pronounced the same way, and *bayou* comes from the Choctaw *bayuk*, meaning "a slow-moving stream." American Indian borrowings account for only a handful of terms, however, most of them referring to local flora, fauna, or topography.

The notion that these differences can render the language incomprehensible to French speakers from Europe is disproven again and again by encounters between speakers of the different varieties, whether in the form of Louisianans visiting Europe or vice versa. It is true that when the language is spoken very quickly, the uninitiated may sometimes have difficulty understanding, particularly if they are not familiar with another nonstandard variety of French. But with some minor effort on the part of both speakers, successful communication is almost always possible. When I arrived in Louisiana speaking the French I had learned living in the Parisian region, for example, I encountered very few misunderstandings (in either direction), though it was immediately clear to everyone that I had learned my French elsewhere.

The final myth—that Louisiana Regional French (i.e., "Cajun French") is Acadian French that has been preserved through isolation—is perhaps the most difficult to eradicate. It is largely due to the widespread use of the ethnic label *Cajun*, which is derived from *Acadian*: Acadian French and other nonstandard varieties possess a phenomenon whereby the [dj] cluster is rendered [dʒ] (in the same way *did you* becomes *didja* in English) before a front vowel, particularly in casual speech. Thus, [akadjɛ̃] > [akadʒɛ̃] > [kadʒɛ̃] becomes English *Cajun*. However, the fact that most of these speakers go by this *ethnic* label, never mind use this language label, is largely a phenomenon of the past thirty to forty years, though it has roots in the social and economic interactions of the post–Civil War period and in social movements beginning in the 1930s. In fact, modern Cajuns are the descendants of several different groups, mainly though not exclusively speakers of nonstandard varieties of French. In some areas—for example Avoyelles Parish—there may be very little *Acadian* ancestry at all, though (white) residents of francophone background nearly uniformly identify as *Cajun*. A brief look at the history of francophone immigration and the developments of the later nineteenth and early twentieth centuries demonstrates that the popular belief that Louisiana Regional French descends from Acadian French belies a much more complex story. While the Acadian immigration is certainly an important element of Louisiana's francophone history, a full understanding requires tracing the origins and probable linguistic history of all the settlers to the state and establishing the Acadians' place in it..

Early Immigration and the Origins of Colonial French

Despite popular misconceptions according to which wealthy Frenchmen migrated to Louisiana early in the colony's history and brought with them upper-crust French culture, Louisiana's earliest francophone colonists were most often drawn from the lower classes.[7] In fact, the French had little success recruiting settlers to the new territory. Extensive propaganda campaigns launched by monopoly owners eager to bring in settlers (most notably Scotsman John Law) had little effect: the metropolitan French population had a thoroughly negative view of Louisiana and was consequently extremely reluctant to relocate to what was viewed as a hot, mosquito-infested, disease-ridden swamp unfit for human habitation. Consequently, the people who sought a new home in Louisiana often did so because they had nothing to lose. The earliest settlers (and indeed the founders of New Orleans) were Canadian (i.e., Quebecois) fur traders and adventurers accompanied by French sailors and a handful of buccaneers from Saint-Domingue (Brasseaux 2005:3). A good number of early colonists were relocated by force as the French, desperate to find settlers, resorted to abductions of "problem populations"— prostitutes, criminals, the homeless, and the like—and sent them packing to the colony. This endeavor, however, lasted only four years (Brasseaux 2005:10). About one hundred German-speaking Alsatian farmers arrived in 1721. Most numerous in the early colony were disgruntled, lower-rung soldiers who had drawn the short straw in being assigned to the Louisiana detail and only somewhat reluctantly decided to stay, enticed by the possibility of landownership.[8]

Canadian, Haitian, and Alsatian contingents aside, even the earliest French settlers to Louisiana were not a homogeneous group. Settlers both civilian and military were drawn from all over France, but the majority came from a crescent-shaped swath incorporating roughly the Ancien Régime provinces of Île-de-France, Brittany, Normandy, Picardy, Champagne, Poitou, Aunis, and Bourgogne (Brasseaux 1987).[9] In fact, one-third were drawn from Brittany and Île-de-France, which broadly corresponds to the Parisian metropolitan region today (see figure 3.1).

Origins are known for only 7 percent of the non-Acadian French settlers arriving during the Spanish colonial period (1760–1800). Within that group, Île-de-France, Brittany, and Normandy are represented, but the bulk came from the southern provinces, notably Guyenne and Provence (see figure 3.2).

But what does all this mean for the linguistic history of Louisiana? First, even the ancestor of French as we know it today was the language of only a small percentage of the French population historically. In fact, France was (and still is) a linguistically diverse place, as the map of historical French linguistic regions (figure 3.3) illustrates. Comparing the linguistic map with those showing the origins of French settlers illustrates that these early settlers represented a very broad range of linguistic variation, though most of the settlers came from the region north of the isogloss bundle separating *Oïl* (northern) from *Oc* (southern) dialects. The dialects spoken in those regions, though sometimes surely only minimally mutually intelligible, were similar in many ways to modern Standard French, itself descended from an *Oïl* dialect, and to each other. Moreover, the social class of these early settlers is not a negligible factor: even those originating in Île-de-France were not likely to be speakers of something that closely resembled modern Standard French despite the dialect's origin in that region. Given the city's

Figure 3.1. Origins of French period settlers. Darker areas indicate regions providing more settlers. Map created by the author from data in Brasseaux 2005:16–17.

Figure 3.2. Origins of Spanish-period French immigrants. Darker areas indicate regions providing more settlers. Map created by the author from data in Brasseaux 2005:16–17.

economic and cultural importance throughout its history, Paris was a draw for people throughout the country, and consequently, the language spoken in the streets of Paris was at any point a mix of dialects from various regions. Anthony Lodge (2004), the leading scholar of the development of French, has explained that the working classes of the Parisian population spoke (and still speak) something we call *le français populaire*, a koiné that has its origins in the constant influx of related though diverse varieties of dialectal French. At the same time, the dialect originating in the Parisian region spread initially to other urban areas, where it was colored by local varieties. Many immigrants from provinces other than Île-de-France were also urban, and

Figure 3.3. French dialect regions. Image courtesy of the Institut géographique national de France.

rural settlers often spent time in port cities, where they would have come into contact with urban varieties of French.

All of this suggests that that most or even many of the settlers of French origin during the French and Spanish colonial periods should not be expected to have imported a language that resembled closely modern Standard French. Rather, they brought with them varieties that, while resembling each other to a great degree, were nonetheless neither Standard (even for the time) nor uniform, though contact with French in urban areas may well have had a leveling effect even before the settlers met in the New World.

It is precisely because of this difference in dialect that the standard-like Plantation Society French should be distinguished terminologically from the nonstandard French spoken by the early colonists—Colonial French. It was into this linguistic mix that the Acadian refugees, themselves speakers of a nonstandard dialect, arrived.

The Acadian Immigration

The Acadians were the descendants of French colonists who had fled their homeland in the early seventeenth century and settled in what is now the Canadian province of Nova Scotia, a land they called *Acadie* (*Acadia* in English). In 1755, at the start of the Seven Years' War (or the French and Indian War), they were banished by the British, who had taken control of the area in the interim, in what came to be known as the Grand Dérangement (the Great Upheaval) for a refusal to pledge allegiance to that Crown. Many were removed to the American colonies, often held there for extended periods of time as virtual prisoners, while others went first to Britain and then to France. Still others made their way to the Caribbean. Along the way, about half their number perished, and it is a testament to the willpower and resourcefulness of the group that so many of the survivors managed to find their way to a distant land and reunite. The Spanish, who received control over the Louisiana Territory via secret treaty at the end of the war, saw the Acadians as potential opposition to British expansion in the New World and began welcoming Acadian settlers to Louisiana in 1765. Most Acadians arrived in two major waves immigration, the first between 1765 and 1770 and the second in 1785 (Brasseaux 1987, 2005).

In contrast to the earliest French settlers in Louisiana, the Acadians were a strikingly homogenous group. A full 70 percent of their ancestors came from the Poitou-Saintonge-Aunis region alone (figure 3.4), and within that group, 70 percent came from within a twenty-mile radius of a single town, Loudun, in the province of Poitou. The languages that contributed to Acadian French were Oïl dialects, meaning that Acadian French arrived with a great deal in common with the French already spoken in Louisiana.

While the Acadian immigration was very important and the group was ethnically strong, it would be a mistake to assume that their French was dominant in the colony or that they became so isolated that theirs was the only one to survive (though they did undergo a brief period of at least social isolation in the early years following their arrival [Brasseaux 2005]). The first point of note regarding the Acadian immigration is their sheer minority

Figure 3.4. Acadian origins. Darker areas indicate regions providing more settlers. Map created by the author from data in Brasseaux 2005:16–17.

in the colony. The first wave of about fifteen hundred Acadians arrived in Louisiana at a time when the colony's population included approximately ten thousand people (slave and free) of Old World origin. Thus, the Acadians represented about 13 percent of the overall population at the time.[10] Twenty years later, when the second wave of Acadians arrived, the population of Louisiana had topped forty thousand, even prior to the influx of about ten thousand francophone refugees from Saint-Domingue in 1809 (Klingler 2003a, 2009). Even if we consider the fact that the first wave of Acadian settlers had produced a new generation, increasing their numbers, it is hard to consider the Acadians anything but a small minority in this sea of immigrants (see table 3.2).

Table 3.2 Louisiana's Population of Old World Origin, 1766 and 1788		
	1766	1788
Free	5,611	19,455
Enslaved	5, 799	23,166
Total	11,410	42,621

Source: Klingler 2009:93

In addition to their numerical inferiority, the Acadians did not wander far from their initial settlements—at least, not in great numbers. In fact, over the years, Acadians showed a remarkable resistance to expansion, preferring to remain near their friends and relatives. Brasseaux (1987:115) notes that "Acadian pioneers consistently ventured only as far as the nearest unoccupied waterfront property."

Despite a reluctance to move, Acadians did not remain isolated for long. The first wave of immigrants settled in areas already settled by other francophones, and Acadians everywhere were joined by other groups of immigrants. Following the Civil War, wealthy planters and nineteenth-century middle-class immigrants, economically ruined by the war, joined the Acadians on the bottom rungs of the socioeconomic ladder. The end result was rampant intermarriage, with half of Acadian marriages taking place with non-Acadians (Brasseaux 1992). The high rate of intermarriage, which produced a strong pressure to assimilate, resulted in a new, hybrid culture. Even in areas where they were numerically superior, Acadians absorbed foreign practices, significantly altering Acadian society (Brasseaux 1992:109).

The Evolution of Cajun and of Louisiana Regional French

Further complicating the situation was Anglo-American settlers' failure to differentiate between francophone groups and application of the term *Cajun* to any poor, rural, white francophone. As Brasseaux (1992:104–5) explains, "The term *Cajun* thus became a socioeconomic classification for the multicultural amalgam of several culturally and linguistically distinct groups." During this period, the francophone ranks were also increased by the assimilation of other anglophone, germanophone, and hispanophone groups, thus rendering entire families of Webers, Griffins, and Torreses, for example, francophone (Brasseaux 1992:39, 106). (Some Gallicized their family names accordingly, so that Weber, for example, became Webre.) The process of Cajunization that characterized the latter half of the nineteenth century continued into the twentieth, receiving further impetus from the insistence on a Cajun identity linked to Acadian ancestry that began in the early twentieth century (Brundage 2000) and that was continued in particular by CODOFIL following its establishment in the late 1960s. Cécyle Trépanier (1991) documented this ongoing phenomenon, noting that people who had previously called themselves *Creole* (a contentious term in itself; in this context, it refers to white, non-Acadian francophones) were now calling themselves Cajun. Shana Walton (2017b), who conducted research in Terrebonne Parish in the late 1980s, notes that almost none of the older people called themselves Cajun at the time; rather, they were *French*. By the time I began research in Louisiana in 2003, the process was all but complete. Though in a few areas (Plaquemines and Natchitoches Parishes, peripheral to the rest of francophone Louisiana) residents still insist on the term *French* as an ethnic identifier, in most areas I have encountered only sporadic acknowledgments among rural white francophones that anyone had ever called themselves something other than *Cajun*. Thus, areas such as Evangeline and Avoyelles that were historically home to very few Acadian immigrants are now known as hubs of Cajun culture (Brasseaux 1987; Klingler 2009).

As people mixed with each other, their dialects mixed, too. Moreover, many of these dialects were so similar to each other that it is very difficult to attribute features of modern Louisiana Regional French to any one source, Acadian or otherwise, and numeric superiority is not an easy predictor of retention of features. Indeed, areas with high concentrations of Acadians often lack features that are found in modern Acadia, and well-known Acadian features appear in regions with few Acadian settlers. For example, the replacement of the phoneme /ʒ/ with /h/ (see chapter 4) is a hallmark of Acadian speech. In Louisiana, it is predominant in lower Lafourche Parish, an area of low Acadian immigration, but entirely absent in more heavily Acadian Lafayette. Of course, most if not all of the features typical of Acadian French were brought to Acadia from France, and settlers did come to Louisiana directly from the same region that provided the bulk of the Acadian immigration. The presence of a hallmark Acadian feature in an area of low Acadian presence cannot be solely attributed to the presence of Acadians, however numerically inferior. Today's Louisiana Regional French may be descended from the nonstandard dialects of the early French immigrants, Acadian French, and the near-Standard French of the later nineteenth-century immigrants in different proportions, depending on the historical demographics of a region. (See Picone 2006 for a discussion of the difficulties of assigning provenience to any one feature.) In fact, distinct regional settlement patterns resulted in the creation of a fairly high degree of (superficial but nonetheless important) variation despite a general uniformity across the state. The primary features that distinguish Louisiana Regional French from modern Standard French—the unrounding of front rounded vowels, the use of *être après faire* and *asteur*, the use of an apical *r* and so on—are widespread in Louisiana French. However, the high degree of superficial variation means that speech can often enable at least a rough identification of a speaker's origins within Louisiana.

VARIATION IN LOUISIANA REGIONAL FRENCH

Variation in Louisiana Regional French is often very localized. For example, several isoglosses (the geographic boundary between two variants) pass through Lafourche Parish roughly in line with the Intracoastal Canal—i.e., between upper Lafourche Parish and lower Lafourche Parish. The importance of the distribution of variants cannot be determined by science, however—there is no objective way to rank the importance of features in making dialectal divisions, though some have put forth some tentative suggestions (e.g., Chambers and Trudgill 1998:98), or any mathematical formula for determining how much variation separates dialects from subdialects (or, for that matter, languages from dialects). Danish, Swedish, and Norwegian are largely intercomprehensible but are considered separate languages; Mandarin and Cantonese are mutually incomprehensible yet are considered dialects of the same language. While it may be clear that English and French are not the same language due to their mutual incomprehensibility, the boundaries between dialects and often even between languages tend to be fuzzy. These divisions are to a large degree political and ideological: the importance of a feature in distinguishing one dialect from another depends on speakers' evaluations of their relative importance—evaluations that are often influenced by nonlinguistic factors. The French language as it is spoken in the Lafourche Basin both shares features with French across the state and is demonstrably distinct from it. In addition, it attests variation along ethnic lines. Whether people emphasize or minimize these differences is subjective and an essential part of the linguistic construction of place.

Chapter 4

FRENCH IN THE
LOWER LAFOURCHE BASIN

The French spoken in Terrebonne-Lafourche is the unique product of the history of the region. The language reflects the presence of both Acadians and other French speakers, and the distribution of variable features across the region reflects the history of contact between those French speakers and their anglophone, germanophone, hispanophone, and American Indian neighbors and the movements of communities across the region over time

It is generally easy to identify a speaker from the lower Lafourche Basin, as a perception exercise I conducted as part of my initial fieldwork demonstrated. The exercise was a verbal guise (see chapter 5) during which I played eight twenty-second recorded passages of speakers from Terrebonne-Lafourche. All the speakers were male; I played clips from one Cajun and one Indian from each of the four towns I targeted while participants listened via noise-blocking headphones. I then asked the listeners to identify the speakers' ethnicity and geographic origin. I encountered a fair degree of bias when I presented the exercise to residents of the lower bayous (see chapter 5). To avoid the problems with local bias, therefore, I took the exercise outside the parish. I first went to the town of Mamou, about 170 miles away on the prairie, and then to Choctaw and Raceland in upper Lafourche Parish. In both places, I asked listeners only to identify the speakers as either Indians or Cajuns, having concluded, following some initial questioning, that asking which town—or even which parish—speakers were from would be too much to ask people not from the immediate vicinity.

The outsiders were terrible at answering the question I posed. The exercise was fraught with interference from ideology: when participants (all of whom were white) had no trouble understanding what speakers were saying, they guessed that the speakers were Cajun. Moreover, participants by default associated French with Cajun identity. In Choctaw, for example, more than one participant announced, "C'est des cadiens [They're Cajuns]" before I had played a single clip. However, they were very good at determining that the speakers were not local. In Mamou, nearly every participant's first comment was that the speakers were not from the area, though in most cases, they could not place the speakers' origin. In upper Lafourche Parish, however, participants not only knew that the speakers were not local but immediately noted, "Ça vient d'en bas du bayou! [That's from down the bayou!]." This is in accordance with the parish-wide survey I conducted with students in 2008, which showed a number of isoglosses passing roughly through the area between Larose and Mathews, towns that separate "up the bayou" from "down the bayou" and divide the parish into two distinct regions.[1] These isoglosses represent variation at all levels.

LEXICAL PARTICULARITIES

Lexical variation—difference in vocabulary—is often more salient to speakers of different dialects than are differences in pronunciation or word formation. Terrebonne-Lafourche can be distinguished from other parishes by several lexical differences; perhaps the most important, given the frequency of its occurrence, is the use of a single interrogative pronoun, *qui*, for both animate and inanimate referents (i.e., *qui* means both "who" and "what"). Most other regions follow Standard French in having separate animate (*qui* [who]) and inanimate (*quoi* [what]) pronouns. Key exceptions to this pattern include Evangeline and Avoyelles Parishes, where *quoi* is attested but inanimate *qui* is predominant (figure 4.1). The use of inanimate *qui* is historically well documented in the French language and appears in other regional dialects of French (Rottet 2004). Rottet (2004) suggests that its use in Terrebonne-Lafourche despite the presence of Acadian immigrants, who would have used *quoi*, may result in part from the influx of non-Acadian francophones to the region (they may have come speaking dialects featuring inanimate *qui*) and in part from the Acadians' quarter-century sojourn en route to Louisiana in regions of France where it was likely still in active use.

Terrebonne-Lafourche also uses *bois* for "tree" instead of the more common (and Standard) *arbre*. This feature patterns regionally in Louisiana: with a few exceptions, *bois* is used to the east of the Atchafalaya River and *arbre* to the west (figure 4.2). Terrebonne-Lafourche, to the east of the Atchafalaya, falls into the expected pattern.

Some minor lexical variation exists within the region itself. In Lafourche, for example, a mosquito is a *moustique*, while in Terrebonne it is a *maringouin*. A slightly different pattern emerges for the terms for "bucket," "boat

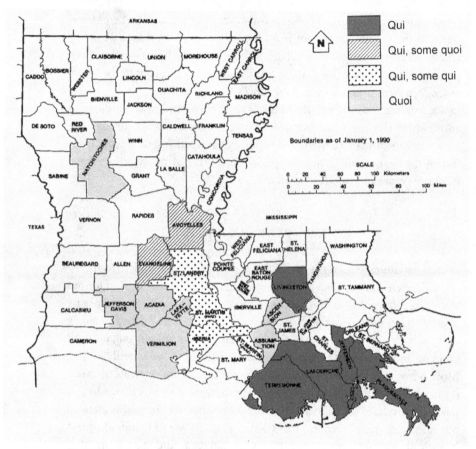

Figure 4.1. Distribution of inanimate pronouns. Map created by the author.

propeller," and "saw": in Terrebonne Parish, the most frequently used terms are *baquet*, *palette*, and *égoïne*. South Lafourche attests only *sciau*, *propelle*, and *scie*, respectively. Upper Lafourche Parish follows the same pattern as Terrebonne; the lines demarcating the exclusive use of *sciau*, *propelle*, and *scie* are part of a bundle of isoglosses that pass through the same rough area and reveal South Lafourche to be an identifiable subregion within Terrebonne-Lafourche. Some of this variation seems to be a recent development. For example, Una Parr's 1940 thesis on the language in Terrebonne Parish lists the term *manche* meaning "lane" (i.e., a small street). However, I was able to elicit the term only from people in lower Lafourche Parish, and when I asked people in Terrebonne directly about it later, the general consensus was that it was a Lafourchism.

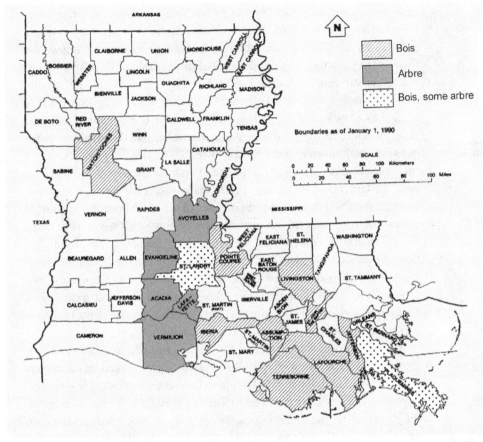

Figure 4.2. Distribution of *bois/arbre*. Map created by the author.

PHONETIC AND PHONOLOGICAL FEATURES

Some of the variation in Terrebonne-Lafourche is highly unstable. The pronunciation of orthographic *oi* as [wɛ] or [ɛ] rather than the standard [wa] in words such as *boîte* (box) or *droite* (right) is one such example. Historically, [wɛ] predates both [wa] and [ɛ]; the latter are the result of sound changes originating in the thirteenth century that took centuries to move to completion; the use of [wa] did not become standard until the late eighteenth century, though it was found in the *français populaire* of Paris in the latter half of the sixteenth century and may have been used by courtiers at that time as well (Ayres-Bennet 2004). The use of [ɛ] appears to have originated in the court and spread to the population at large by the second half of the

seventeenth century (Ayres-Bennet 2004). North American French varieties, including Métis French (Papen 2004; Rhodes 2009), Quebecois, and
Acadian French (Flikeid 1984; Walker 1984; Poirier 1928) are well known for
their use of [wɛ], though most if not all of these communities also use [wa]
and other variants (including [wɛ], [ɛ], [wɔ]). In Terrebonne-Lafourche,
however, much as was the case with the palatalization of velar stops, [wɛ] is
limited to a few lexical items. These include *droite* (right), *boîte* (box), and
soye (pronounced [swɛj]), the local subjunctive form of the verb *être* (to be;
cf. *sois/t* [swa]). Even in these words, however, its use is not categorical, and
the same speaker may alternate the use of both forms. As it was difficult to
elicit the terms in spontaneous speech, there is too little data to determine
whether the alternation is governed by social context. Given the very small
number of words affected by the phenomenon, however, it seems more likely
that this is simply a case of competing forms, comparable to the pronunciation of *caramel* in English as documented by the Harvard Dialect Survey
(Vaux and Golder 2003), for which 17.26 percent of speakers claim to use two
pronunciations interchangeably. Possibly in this case it is due to the short
time depth involved: enough time has not elapsed to permit all the variation
present in the different source dialects to stabilize.

The most diagnostic feature of the lower Lafourche is most definitely
governed by social context; it is also a feature whose spread was likely ar-
rested by the shift to English. In Louisiana, the feature that instantly identifies
a speaker as originating in the lower Lafourche basin is the replacement of
the phoneme /ʒ/ (the sound represented by *g* or *j* in such words as *jamais* or
rouge) with /h/. The only exception to this geographic restriction is its sporadic presence in Grand Isle, a half hour's drive away from Golden Meadow.
This is not much of an exception, however: while Grand Isle is technically
located in Jefferson Parish, it has very close historic and current ties with
lower Lafourche Parish. In any case, while the phenomenon is found there,
it is rare, and it would be hard to mistake a speaker from Grand Isle for
someone from lower Lafourche. Grand Isle speakers use a number of lexical items not found in Lafourche, and vice versa. For example, a wall is an
entourage in Lafourche but a *muraille* in Grand Isle, and I have been unable
to find anyone who recognizes the term *vire-veaux* (i.e., *virevolte* [about-face,
spin], but in this case, a passage through a fence in place of a gate consisting of a series of sharp turns that permitted the passage of people but not
cattle due to the length of the latter) in Lafourche. Most notably, Grand Isle
speakers use the Standard French uvular *r* (for a possible explanation of this
phenomenon, see Picone 2014, 2015). This is the most obvious difference to

locals: when asked where people speak French differently, people from lower Lafourche most often point to Grand Isle and suggest that people there clear their throats incessantly.

The alternation is related to stylistic choice in response to social context or register (Carmichael 2007; Salmon 2007; Dajko 2009). Contexts in which one is expected to speak well (for example, a court proceeding or a classroom, to different degrees) promote the use of standard speech (i.e., the most "correct" speech available to a speaker). Familiar contexts such as casual conversations with family and friends promote the use of vernacular (i.e., nonstandard) forms. In linguistic interviews, following methodology pioneered by Labov (1984), one way of attempting to access the most "correct" version of speech is to focus the interviewee's attention on their speech; conversely, in order to access a speaker's most vernacular speech, the interviewer attempts to draw their attention away from their speech.[2] Asking interviewees to read a list of words or a short passage (or both) is assumed to draw attention to speech; the vernacular is elicited by drawing attention instead to the topics of conversation, with topics that evoke an emotional reaction, like a dramatic story of some sort, providing the most vernacular, as the interviewee is emotionally engaged in the topic. The majority of Louisiana francophones are unfamiliar with French orthography and consequently cannot read French; accordingly, I replaced reading exercises with an exercise in which I asked them to translate English sentences targeting the features I had chosen for examination into French. I targeted informal registers by casually discussing the interviewee's life history and the local culture, asking in particular about dramatic events such as hurricanes.

In Lafourche, /ʒ/ is replaced by /h/ more often in casual speech than in careful speech (Carmichael 2007; Salmon 2007; Dajko 2009). Moreover, within the lower Lafourche basin, the lenition of /ʒ/ is not only linked to style; its use also varies by both by geography and ethnicity. The core region for the phenomenon is the Larose–Golden Meadow corridor along Bayou Lafourche, where I found it attested between 33 and 48 percent of the time in the formal translation exercise I administered (i.e., careful speech); in casual speech (a general discussion of whatever topics interested the interviewee), this number rises to about 60 percent. The change in the pronunciation of /ʒ/ when moving between casual and careful contexts has been previously addressed by Salmon (2007) for upper Lafourche Parish and Carmichael (2007) for the Indian population in Pointe aux Chênes; both authors conclude, given the same pattern, that speakers are still able to alter their speech in some way, however subconsciously, to respond to changes in context. This

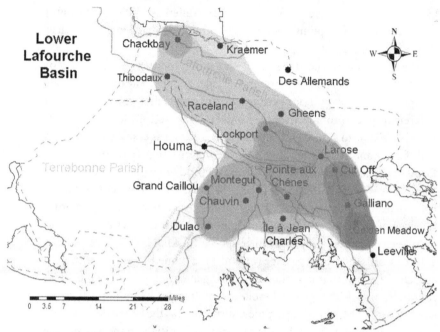

Figure 4.3. Distribution of /h/ substitution in Terrebonne-Lafourche. Houma is not included simply because no data exists for the city. No data exists for Montegut or Chauvin, either; however, it was impractical to shade around those two towns, so they are included in the shaded area for the towns to the west, Grand Caillou and Dulac, and that inclusion reflects an assumption rather than fact. Map created by the author.

is particularly notable given the collapse in registers favoring the informal that has occurred as a result of language shift—for example, I have only sporadically encountered people who use the polite/formal second-person singular pronoun *vous*, and then very rarely with the expected *-ez* morphology on the verb. My own data, the numbers for the Pointe aux Chien Indian community corresponding nearly exactly to Carmichael's, shows that context is the motivating factor governing the replacement of /ʒ/ with /h/ for all areas in Terrebonne-Lafourche that attest it with the exception of the Île à Jean Charles.

The phenomenon peters out fairly rapidly as one leaves lower Bayou Lafourche (figure 4.3). Residents of Pointe aux Chênes only use /h/ roughly 25 percent of the time in careful speech, and those in Dulac use it even less (15 percent), though in casual speech the rates jump to 37 percent and 40 percent, respectively. The Île à Jean Charles, the most physically isolated of the research communities, has the lowest frequency of all the lower bayou communities, at roughly 10 percent in careful contexts and 12 percent in

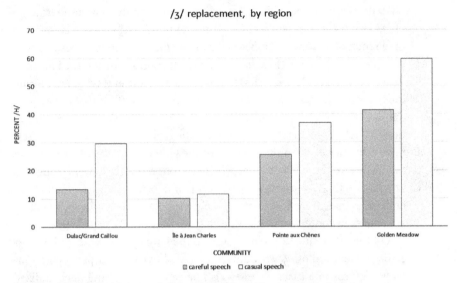

Figure 4.4. /h/ replacement by community in Terrebonne-Lafourche. Image created by the author.

casual contexts (figure 4.4). This lack of stylistic difference suggests that it is in fact a recent innovation in that community. Going north, the same decline in use is apparent. Communities in central Lafourche Parish are similar in rate to Pointe aux Chênes, but heading north into the vicinity of Thibodaux, in upper Lafourche Parish, the feature is already all but absent, appearing less than 1 percent of the time in careful speech. Across the lower bayous, it can function as an ethnic marker to some degree, as Cajuns are more likely than Indians to use it except in the community of Dulac/Grand Caillou, where the situation is reversed.

The /ʒ/ may be replaced in any position and without phonological constraint. Thus we find pairs such as *jambon/hambon* [ʒɑ̃bɔ̃]/[hɑ̃bɔ̃] (ham), *manger/manher* [mɑ̃ʒe]/[mɑ̃he] (to eat), and *piège/pièhe* [pjɛʒ]/[pjɛh] (trap). Particularly along lower Bayou Lafourche (i.e., in the core region), the change to /h/ may also affect the phonemes /z/, /s/, and occasionally /ʃ/, though less frequently than it affects /ʒ/. (The most frequently replaced of the secondary phonemes is the voiced /z/, with /s/ and /ʃ/, in that order, falling far behind in frequency.) Thus, *nous-autres* becomes *nouh-autres* and *maison* becomes *maihon*. This aspect of the phenomenon has not been examined in previous studies; my data suggests that it follows the same pattern as the /ʒ/ replacement (i.e., that it is used more often in casual speech). Its inclusion in the phenomenon also appears to be unique in the francophone world, as no other descriptions include any sound other than /ʒ/, though the alternation of /ʒ/ and /h/ is otherwise well documented internationally; it is found in the French of southwest France (Horiot and Gauthier 1995; Hewson 2000) and in Acadian French (Flikeid 1984, 2005; Hewson 2000) and the French of various regions in Quebec (Hewson 2000; Charbonneau 1957; Boissonneault 1999), though in many if not all of those places it is not /h/ that replaces /ʒ/ but rather the velar fricative /x/ (the final sound in the German pronunciation of Ba**ch**). It seems that in Lafourche the replacement sound was simply moved further back, to the position of the glottal fricative /h/. In any case, the retention in the lower Lafourche basin but not elsewhere in Louisiana of a feature that is almost certainly Acadian in origin (for this population) underlines that numeric superiority is not a simple predictor for the retention of features in contexts of dialect mixing.

On the other side of the region from the Larose–Golden Meadow corridor, the lenition of /ʒ/ encounters an added twist that is particularly interesting. In some cases, /ʒ/ may also be replaced with the depalatalized [z] (or [s] depending on phonological context; the latter occurs most often before voiceless consonants, and the former generally in all other contexts) rather than with /h/. So, *je crois* may become *s'crois*, and *je l'aime* becomes *ze l'aime*. The phenomenon is almost exclusively limited to the first-person singular subject pronoun *je*. While attested across the region, this alternation is concentrated in the Dulac/Grand Caillou area and is only a very minor phenomenon outside of it, occurring less than 5 percent of the time on the Island and less than 2 percent of the time in Pointe aux Chênes and Golden Meadow. Two things make it particularly interesting. The first is that the replacement of /ʒ/ with /z/ is part of a three-tiered style-based variation: while /h/ replaces /ʒ/ in casual speech, /z/ is a feature of careful speech. This is counterintuitive: /ʒ/ is the Standard French variant and the more prestigious variant even in places that feature /h/ replacement, including the Lafourche Basin. Consequently, for another sound to represent a prestige form is unexpected. Moreover, since both /h/ and /ʒ/ exist, with /ʒ/ more prestigious than /h/, that /z/ should appear more often in careful contexts suggests that /z/ occupies the top tier in the hierarchy, given that while /ʒ/ is still more frequently used in the translation exercise it occurs most often in that context. It occurs only sporadically in words other than the pronoun *je*; however, it is likely related to the replacement of dental fricatives with palatal fricatives (and vice versa) in other words across Louisiana. For example, it is common across the state for *chemise* to be pronounced *chemige*, and *chez* is almost always homophonous with *c'est*

[se]. Within Terrebonne-Lafourche, it appears most often in the word *chose* (pronounced *choge*), though I have also encountered, sporadically, *chaige* (*chaise*), *chauvage* (*sauvage*), and *Etage-Unis* (*Etats-Unis*). I have come across one written representation of *choge* in a letter written by a local man in the early twentieth century (Allen 1986:19). It is documented in the literature on Louisiana French varieties, notably by Alcée Fortier (1891:88) who observed, regarding what he called the Acadian French of Louisiana, that "j [by which he meant /ʒ/] is sometimes z: Zozé for Joseph" and that "z is sometimes replaced by j: Jénon for Zénon." The variation is found in Louisiana Creole as well. Fortier (1884–85:xlii) described the same phenomenon for that variety, listing the example *changé* > *chanzé*, and he further described the alternation between [ʃ] and [s], giving *songé* > *chongé* by way of example. Neumann-Holzschuh (1987) also documents the phenomenon, providing such examples as *jonglé/zonglé* (to think). It appears to be lexically fixed, however, in most cases; an interviewee from Ville Platte who had studied extensively in France and Quebec explained to me that she knew the "correct" form was *chemise* but that her natural instinct was to say *chemige* even in a careful context. In Dulac/Grand Caillou, it is robustly variable within the first-person pronoun.

I have found documentation of the existence of a similar three-tiered hierarchy in the literature only once, though the comparison is not perfect, and to the best of my knowledge, this is the only place in the francophone world where such a phenomenon has been documented.[3] It seems likely that it is the result of a sort of stylistic reallocation (Trudgill 1985, 1986) that occurs in the process of koinéization. That is, the speakers who attest the alternation may be in the process of streamlining a very complex system of phonemes and allophones produced by the meeting of multiple dialects with similar but not identical systems.

ETHNIC-BASED VARIATION IN TERREBONNE-LAFOURCHE

The three-tiered hierarchy is one of three clear markers of ethnic identity in Terrebonne-Lafourche, separating the speech of Indians from that of Cajuns. In this way, the language is a repository of the history of community migration and interaction in the area. Equally interesting is that the phenomenon seems to have originated in the Indian community. Figures 4.5 and 4.6 show the rate of /z/ replacement in Dulac and particularly for Cajuns and Indians in the first-person subject pronoun.

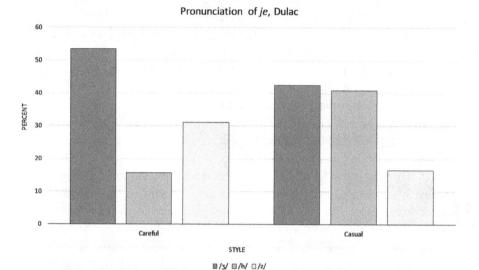

Figure 4.5. Alternative of initial sound in 1sg subject pronoun, Dulac. Image created by the author.

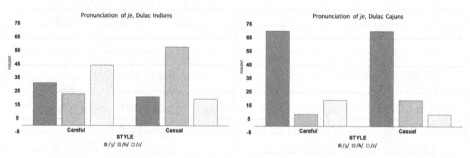

Figure 4.6. /z/ replacement in 1sg subject pronoun by ethnicity, Dulac. Image created by the author.

While in casual speech, the two groups' use of /z/ differs by only 7 percent, in careful speech, the Indians use /z/ at over twice the rate that Cajuns do. Moreover, Cajuns have a strong preference for /ʒ/ regardless of context; Indians show the expected style-based adjustment in their use of /ʒ/, but in no context is it the most frequently used, and most of the difference between contexts can be attributed to the shift from /z/ to /h/. It seems, then, that when pronouncing *je*, /z/ is considered more prestigious by Indians and that Indians are using it to express their identity in careful speech rather than in casual speech.

Variation in the language according to ethnicity has rarely been investigated by researchers. Indeed, scholars and government agencies have long claimed that the Indians have fully acculturated to the lifestyles of their Cajun neighbors, including the adoption of a French indistinguishable from that spoken by them. It is perfectly understandable that a casual listener might believe this to be the case: with only one exception, the variation results from variance in rate of use that emerges only under close statistical scrutiny rather than from categorical difference. In an ethnographic report for the Université de Laval's Projet Louisiane, Larouche (1981:7) is representative in his claim that "francophone de part et d'autre, l'Indien manie le langage de la même façon et avec le même accent que le Cadjin [The Indians are thoroughly francophone across the region and use the language in the same way and with the same accent as the Cajuns]." While several authors have documented the French of the area (Guilbeau 1936, 1950; Lecompte 1962, 1967; Papen and Rottet 1997; Parr 1940; Oukada 1977), these studies were conducted without using the ethnic affiliation of those studied as a variable to be investigated. Of particular note, Oukada (1977) mentions the ethnic groups present in Lafourche but deliberately chooses to collect data only from self-identified (white) Cajuns. Guilbeau (1950), meanwhile, asserts that all the ethnic groups in Lafourche Parish speak the same French. As Guilbeau was himself a native of the parish and as a native speaker of French used himself as a primary informant, this statement appears to be based on informal observation or personal opinion and not on empirical evidence. This is particularly striking, since Guilbeau was somewhat ahead of his time, carefully collecting data from all regions of the parish with an eye toward considering variation. He certainly notes the existence of Indian communities in the southern end of the parish, but none of his informants were from that community, and there is no indication of the racial or ethnic affiliation of the informants he worked with.

My dissertation (Dajko 2009) thus represents the first large-scale systematic study of the area with the goal of documenting ethnic-based variation. While Rottet (1995, 2001) takes ethnicity into account in his examination of language death, that study did not focus specifically on ethnicity. Nonetheless, he identified six items unique to Indian speech in Terrebonne-Lafourche and one that was more commonly attested in Indian speech than it is in Cajun. Using Rottet's features and observations from preliminary interviews as a starting point, I compared the speech of Indians and Cajuns extensively and not only added nuance to his findings but also identified several additional features that are deployed by speakers to express their ethnicity. The strongest indicators, in addition to the three-way alternation of /ʒ/ with /h/ and /z/, are the use of the masculine third-person singular subject pronoun *il* to indicate a referent of either gender and progressive nasalization.

Table 4.1. Use of 3sg Feminine *il* across Communities		
Community	3sg *il*, Careful Speech*	3sg *il*, Casual Speech
Dulac/Grand Caillou	9.5%	6%
Île à Jean Charles	39%	47%
Pointe aux Chênes	55%	60%
Golden Meadow	1.5%	6%

* These numbers are taken from the translation exercise including all instances of potential mistranslation. When those instances are removed, the Island slips into first place and Pointe aux Chênes into second. The change in pattern suggests that most possible mistranslations are not in fact mistranslations.

The use of the third-person singular masculine subject pronoun *il* (he) to refer to both males and females emerged from the data, contra Rottet's findings, as the only shibboleth of Indian speech. Thus, *My daughter would buy that car if **she** had enough money* is translated as *Ma fille ajèterait ce char si **il** aurait assez d'argent*, and **She** *bakes bread on Mondays* is *Il faisait du pain le lundi*.[4] In Louisiana we would expect instead the use of *alle* or *elle* (*elle* being shared with Standard French), both of which are also found both in Cajun and Indian speech, with *alle* the more common variant, *elle* generally reserved for the objective case (i.e., *her*). This was the feature most often cited by Rottet's interviewees, though it was only mentioned twice in my interviews; most people preferred to declare, "C'est proche tout pareil [It's pretty much (literally *almost all*) the same]." It is exclusive to Indian speakers and most frequent in Pointe-aux-Chênes and the Island. It is unclear that style is a factor in this case; while the numbers show a decline in the use of feminine *il* in careful speech across the board, it may result from the nature of the sentences used to elicit the feature or missed counts in free speech due to ambiguity as to the gender of a referent (see table 4.1).

In any case, though it is exclusive to Indian speech, the feature is not common—the overall rate for the Indian community across all four towns is only 27 percent in the translation exercise and 30 percent in casual speech. It is also not evenly distributed across the population. Even in Pointe aux Chênes and on the Island, where it occurs most frequently, many speakers do not use it at all. Interestingly, there is also a slight upward trend in the use of the feature as age declines: younger people are slightly more likely to use it than are older people, suggesting that it is an innovation rather than a retention.

The last feature that is particularly indicative of Indian speech is the progressive nasalization of the midfront vowels /e/ and /ɛ/ so that, for example,

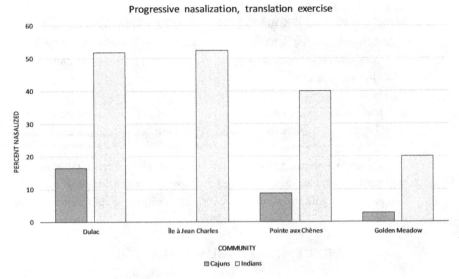

Figure 4.7. Progressive nasalization of /e/. Image created by the author.

mais [mɛ] or [me] (but) sounds like *main* [mɛ̃] (hand) due to continued nasalization following the nasal /m/. The feature also patterns geographically, with the highest rate in Dulac/Grand Caillou and the lowest in Golden Meadow (figure 4.7). The difference between Cajun and Indian rates of nasalization remains stark regardless of which town is under examination, however.[5]

Other features that pattern with Indian identity are less conspicuous. Like progressive nasalization and the three-way hierarchy associated with the lenition of /ʒ/, they are used by both communities but at different rates, and the difference is decidedly less significant than it is for progressive nasalization. These include, but are not limited to:

1. The agglutination of /l/ to verbs beginning with a vowel, presumably due to reanalysis of the frequent use of an elided *le* or *la*: one often hears *je l'aime* (I love it) or *je l'aime, [item]*, rather than *j'aime le/la [item]* (I love the [item]). Consequently, the verb is understood to be *laimer* rather than *aimer*, resulting in *je laime [item]*.

2. The use of a null pronoun (i.e., no overt pronoun) in both first-person and third-person singular contexts. This occurs only when the subject has been previously established and/or the pronoun is consequently obvious to the listener. For example:

(4.1)

R2: Tu peux nous dire des contes?	R2: *Can you tell us some stories?*
L1: Oh *no!* {laughs} Peux pas dire	L1: *Oh, no! {laughs}. Can't tell*
des contes; je connais pas. [cf.	*stories; I don't know any*
Je peux pas dire . . .]	

3. A preference for the standard [wa] in place of [wɛ] when pronouncing orthographic *oi: boîte* [bwat] (box) is pronounced *bwette* [bwɛt].

All of the above are used more often by Indian speakers. Several other features pattern in complex ways, and ethnicity may intersect with geography: the indigenous community shows more regional variation than does the Cajun.

INDIGENOUS LANGUAGE(S) AND FRENCH
AS AN INDIAN LANGUAGE

Equally important to the discussion of variation patterning along ethnic lines is that the expression of Indian identity via language is accomplished not only by the strategic deployment of variable features but also via the use of the French language. Before broaching that topic, however, it is germane to discuss the presence and importance of indigenous language(s). The only documentation of an indigenous language in the lower Lafourche Basin comes to us from Swanton (1911), who visited lower Pointe aux Chênes in 1906 and collected a list of seventy-five words and three sentences in a language he called Houma. The nature of that list has been disputed over the years (see Brown and Hardy 2000), but that matter does not fall within the scope of this work. Rather, I wish to address the matter of living indigenous language on the bayou. To date, with the exception of recent immigrants and occasionally their children, I have not met anyone in the lower bayous who speaks, even poorly, an indigenous language, though rumors abound that such people exist. I followed all the leads I got in that regard, but to no avail: the alleged speaker would disavow any knowledge of such a language. Moreover, the only borrowings I encountered were the same borrowings that appear in the speech of Cajuns (see chapter 3).[6] Several older residents spoke of long-dead parents and grandparents, who had been fluent speakers of an indigenous language, but only one person was able to give me much by way of example of their speech. With much difficulty, he provided me with six terms and their translations, which he said his grandfather had used:

[isa]   "deer"
[nija] or [naja]   "bear"
[oki]   "water"
[saba]   "pig" (further research reveals this is more likely "horse")
[fufu]   "grandfather"
[kyl]   "move, get out"

I obtained one other word from a resident of the Île à Jean Charles:[ʧakalata] "hawk."[7]

A quick comparison with potential sources given the region's history (Chitimacha, Choctaw, and Mobilian Jargon) reveals the words to be clearly Muskogean in nature, with the exception of [kyl] and [ʧakalata], whose origins are indeterminate (meaning I have not found anything that resembles them anywhere to date.)[8] The exact linguistic affiliation of the clearly Muskogean words is unclear, since as a unit they are not a perfect match for any one known language, including Swanton's list. My source for the first six words said he thought that his grandfather had called the language Choctaw, which given the historic semantics of that term (Kniffen, Gregory, and Stokes 1987:95) is not necessarily specific, and my source for the seventh word gave only "Indian" as the origin. Given the generational distance between my sources and those who had used the words and the mistranslation of [saba], it is likely that they may have been slightly misremembered. In any case, they are not words currently used by the community.

It is possible, of course, that any remnants of an indigenous language were considered private cultural capital not to be shared by the few remaining speakers and that they consequently lied in denying any knowledge of such to me, an outsider; it is also possible that I simply did not find the few remaining users of the words/language. In any case, even the former presence of indigenous languages is certainly important to Indian identity. But it is also clear that French, a European import, has become an Indian language as much as it is a Cajun one—and in some ways more so, given that the Indian population retained French a generation longer than did Cajuns. Because segregation resulted in a lack of schooling for indigenous children and therefore less exposure to the strong pressure to abandon French in favor of English, the youngest speakers of French among the Indian population are on average twenty years younger than their Cajun counterparts, and indeed, the handful of monolingual speakers I am aware of are all members of the Indian community. Larouche (1981:7) makes a note of this when, commenting on the use of French among Indians, he remarks, "D'ailleurs le français est devenue *sa* langage au point où nombreux sont ceux qui ne connaissent pas l'anglais avant de fréquenter l'école (chose rarissime chez le Cadjin) [Furthermore, French has become *their* [i.e., the Indians'] language to the point that many do not speak English prior to their entry into the school system (a rare occurrence among Cajuns]." French is so closely associated with Indian identity that the compulsory teaching of English imposed on all francophone groups across the state was perceived by Indians in Terrebonne-Lafourche as an attempt to force Indians to assimilate to non-Indian culture (Pointe au Chien Indian Tribe 2005:9). This was expressed to me directly by a community member during a casual conversation. As we were chatting, waiting for the monthly meeting to begin, one of my interlocutors declared that Indian identity was affected by the loss of French: "When they told us we couldn't speak French in the schools, it was an attack on being Indian." Christine Verdin, a member of the Pointe au Chien Indian Tribe, presented a talk on the loss of French rather than another ancestral language at a 2013 Tulane University conference on Louisiana's indigenous languages. That this association of French with Indian identity is shared by other groups in the area is also clear (as is its importance to other indigenous groups around the state, if the response from the crowd at the conference is any indication).[9]

The retention of the language for an additional generation has resulted in the use of French as a sort of shibboleth of Indian identity despite its association with Cajun identity as well: everybody knows that Indians are more likely to still speak French than are Cajuns. When I asked residents where to find French speakers to interview while conducting a pilot study in 2006, I most often received the names of Indian residents, and I was invariably told to go down the bayou: "There's nothing but Indians down there, and they *all* speak French!"

A May 2006 conversation in Dulac illustrates this point. While I was interviewing an elderly woman, our discussion turned to her children and whether they spoke French. An Indian woman in her mid-forties sitting nearby (identified here as L1) joined the discussion to offer her insight on the issue:

(4.2)

L1: A—a lot of times 'cause you know I—I used to work as a school substitute teacher? A lot of the younger generation didn't want to learn French. You know, because automatically when you talk French, you was known as Indian. And a lot of them didn't want to be recognized as Indian.

R1: You were known as Indian as opposed to—why not also—I mean,
why wouldn't they— I mean why would they assume Indian, and
not Cajun?

L1: That's how it was down here....

R1: So it was just by speaking French that they would be known as
Indian.

L1: Yeah. You know when I went to school, when I started. And if you
spoke French you was automatically uh, you know, the Indian. I
don't know about her [the interviewee's] generation but that's how it
was, you know, for me.

Somewhat ironically, then, while the abandonment of indigenous languages
in favor of French still marks these groups in the eyes of outside organiza-
tions as no longer possessing distinct features indicative of Indian identity,
within Terrebonne-Lafourche, the active use of French is very much a marker
of indigeneity to the local population, both Indian and Cajun. However, while
ethnic identities are most often invoked on a daily basis, French is still very
much associated with Cajun identity as well; it is most accurate to say that
both Cajun and Indian identities are bayou-based identities.

AWARENESS AND THE IMPORTANCE OF
THE LOCAL DIALECT TO IDENTITY

Residents of Terrebonne-Lafourche are generally aware that their language
is distinct from that spoken in other regions, including other parts of Louisiana. Even so, ideology—people's subjective beliefs about language (see
chapter 5)—plays a strong role here regardless. The feature that is exclusive
to the area and that is most striking to anyone not from there—the lenition
of /ʒ/ to /h/—is not one to which speakers attribute the most importance.
When asked to describe how French varies around the state, they most often
point to the most amusing examples of lexical or semantic variation, but
they are also quite aware, for example, that in Grand Isle the uvular *r* is
used. Awareness of the patterning of /ʒ/ lenition varies, with most people
unaware of the reasons for which they do it. When I asked people why they
sometimes pronounced *jamais* with /ʒ/ and sometimes with /h/, the most
common response was confusion: "What do you mean?"

"You know, sometimes you pronounce it *jamais* and sometimes it's *ha-
mais*," I would repeat, dragging out the initial sounds this time. Very often

that only resulted in a shrug or an "I don't know," but on occasion people would come up with an explanation on the spot. One interviewee, for example, decided that it was "because of masculine and feminine." This is, of course, entirely impossible since *jamais* (never) is neither a noun nor an adjective and therefore is not subject to grammatical gender agreement, as would be the case with, for example, a color term such as *vert* (green) which is *vert* [vɛr] when it modifies a masculine noun (e.g., *le bois vert* "the green tree") but *verte* [vɛrt] in the feminine (e.g., *la maison verte* "the green house").

When I asked people in Terrebonne Parish to describe how French was different (by their own allegation) in Lafourche during my early visits to the region, several supplied "they pronounce *jambe* as *hambe*" by way of example, seemingly ignoring the fact that they themselves alternate /ʒ/ and /h/: as a result of such conversations, the word *leg* was in one of my translation exercise sentences ("A wasp bit me on the leg last night"). In Lafourche, 74 percent of respondents produced *hambe*; in Terrebonne, none did. This suggests that some of the attestations of /h/ are in fact not variation; rather, the phenomenon is lexically fixed. For some speakers, certain words—like *jambe*— are excluded from the variation, and the words this affects vary across parish lines. My data shows that in Terrebonne the alternation is most robust in the first-person singular *je*; this is confirmed by Carmichael (2007) for the Pointe au Chien Indian Tribe and is in keeping with the pattern found other places in which the phenomenon occurs (Boissonneault 1999). In lower Lafourche, however, the nonpronominal words are more variable, as shown by the translation exercise, which elicited largely the same lexical items across participants.

In any case, many speakers are well aware of the phenomenon, even if they do not know why they do it or how it patterns. For example, I received a text message from Roland, my fieldwork assistant, laughing that his father and aunt were arguing over whether men used /h/ more frequently than women did. (I responded, unhelpfully and possibly only throwing fuel onto the fire, that they were both right: men do use it more often than women do [in Lafourche Parish but not in Terrebonne], but the difference, though statistically significant, is only 5 percent, so does it matter?) On another occasion, a friend from Golden Meadow reported that his grandmother was not only aware of it but insisted that /h/ was the prestigious variant and the rest of the state was simply pronouncing things wrong. This last response is key: residents of Terrebonne-Lafourche are generally proud of their local dialect. This is true whether they feel it is prestigious or whether they revel in the covert prestige (i.e., group belonging, insider status) that speaking a nonstandard dialect may bring.

Despite their frequent disclaimer that they did not speak "the *real* French" or "pure French," it seemed to be precisely *because* of this that residents linked the language very closely to their own identity. This became clear when participants were asked what kind of French they felt should be taught in the schools. While some people felt that any French was better than none, they generally agreed that local French would be preferable. Some interviewees went so far as to complain that children encounter outsider French in schools today (when they receive instruction in French at all). Even those who said it did not matter which variety was taught often clarified that this was because once the kids spoke some kind of French, they could use it as a springboard to learn local French. The ultimate goal, in any case, was local French.

One interviewee was already concerned about the intrusion of features from other languages, including other Frenches:

(4.3)

Le français que mon père et eusse avait appris, c'était le vieux français parisien. Ça fait qu'on, on parlait ce français-là pour des années, et quand ce que, uh, le monde a commencé à venir, uh, du Canada et de la France, et, um eusse nous attendait parler, eux voulait tous savoir, "Comment t'appris ce uh à parler ce français-là? Parce-que" eux dit, "ça c'est le vieux français." Et, uh, ça fait que parce-qu'on sortait pas de trop, il était préservé pour beaucoup des années. Maintenant c'est plus comme ça, parce-qu'avec le télévision le *radio* et, uh, les automobiles, pour le monde sortir étou on se mêle plein, et, uh, notre français a été mêlé avec les autres langues.

The French that my father and them learned, it was the old Parisian French. So, we spoke that French for years, and when people started to come from Canada and from France, they heard us speaking and they all wanted to know, "How did you learn to speak French like that? Because," they said, "that's old French." And, uh, it was because we didn't get out much, it was preserved for years. Now it's not like that anymore, because with the television and the radio and automobiles, people get out more and we mix a lot, and our French was mixed with other languages.

(Indian, île à Jean Charles)

The fact that the speaker himself uses standard terms like *maintenant* (rather than *asteur*) or *automobiles* (cf. *chars*), though both terms are rarely heard in Louisiana including on the bayou, illustrates the point he is making. This possible intrusion of standard French is not taken lightly. While others were less concerned that their own language was already polluted, they were very concerned that their grandchildren would learn the wrong variety at school:

(4.4)

R1: Asteur on a, on montre aux enfants à l'école à parler le français non?

R1: Today they're teaching kids French in school, aren't they?

L1: {shakes head}

L1: {Shakes her head}

R1: Huh

R1: Huh

L1: Ina pas, ina pas un vrai tas qui prend français ici, mais je sais pas cofaire. Ina pas un tas—ina pas un tas d'enfants qui prend le français

L1: There aren't, there aren't a lot who take French here, but I don't know why. There aren't a lot— there aren't a lot of kids who take French.

LD: Et le français, apprend à l'école, c'est c'est français qu'eusse a dans le livre, mais c'est pas pareil comme nous-autres on parle.

LD: And the French you learn at school, it's, it's the French they have in books, but it's not the same as what we speak.

R1: Ouais

R1: Yeah.

R2: Mmm-hmm

R2: Mmm-hmm

L1: J—ai um, Aaron Aaron avait commencé u—un de mes mes petits enfants.

L1: I have, um, Aaron, Aaron started uh—one of my grandchildren.

R2: Mmm-hmm

R2: Mmm-hmm

L1: Avec le français, et, a menu icitte il m'a demandé à mon voir qui ce que ça c'était parce que c'était pas (le) pareil, c'est pas un pareil, langage

L1: With French, and, he came here and he asked me to come see what something was because it wasn't the same, it's not the same language.

R2: Mmm-hmm.

R2: Mmm-hmm.

L1: Le français dans le livre d'école et le français que nous-autres on XX pareil (après parler?)

L1: The French that's in the school-books and the French that we're speaking.

(Cajun, Pointe aux Chênes)

On another occasion I was present for a conversation at the Chêne à Caouenne (Caouenne's Oak; Caouenne was the nickname of the owner of the yard where the oak is located), a local hangout for (predominantly male) francophones in lower Lafourche Parish. When others in the group were not in the mood for a serious conversation, one very earnest and vocal resident became annoyed. "Mon, j'après dire de quoi important! [Me, I'm trying to say something important!]," he admonished them, becoming increasingly irate as his friends made light of everything he said regarding language death and the teaching of other Frenches in school. He continued,

(4.5)

Le jeune monde va pas regarder comment ce qu'on parle. Eusse va parler en français, mais tu vas pas les comprendre. Nous-autres, on va se comprendre entre nous-autres on se comprend, mais eusse, eusse va pas parle—eusse eusse va le—eusse va le parler, mais eusse vas se parler une différent manière que nous-autres on se parle. Tu comprends qui j'après te dire hein? Eusse, eusse va parler un vrai français, nous-autres on parle le français cadien, mêlé avec le français et anglais.

The young people aren't going to listen to how we speak. They're going to speak French, but you won't be able to understand them. Us, we'll understand each other, but them, they're going to speak it, but they'll speak it differently than we do. You understand what I'm saying, eh? Them, they're going to speak a real French, we speak Cajun French, mixed with French and English.

(Cajun, Golden Meadow)

At this point the conversation broke down into a discussion of the intercomprehensibility of local French with various other French varieties. Still, the encounter was particularly illustrative of interviewees' consistent insistence that schools should teach local French. The language identifies them as authentic members of the community; no other dialect of the language could perform this role.

This connection is not always readily understood. While conducting the verbal guise test, I brought a friend who was visiting from out of state along with me one day, and he was perplexed about why people would be so upset that French was dying in Louisiana, asking, "Don't they know it's still spoken by millions of people in France?"

By the end of the day, though he had heard very little French—he had instead spent the day with people who had only been asked to listen to sound clips and who often spoke English during the exercise—he had come to understand. That French may be thriving in France is nice but is ultimately irrelevant to people in Terrebonne and Lafourche Parishes. Their own French matters because it is a repository of their history. It brings to life the voices of ancestral populations and demonstrates the movements of people that led to the present.

THE CONNECTION OF FRENCH
TO BAYOU IDENTITY

Perception

Obtaining a full understanding of linguistic variation requires determining not only "objective" linguistic facts but also people's subjective understandings of the ways in which they (and others) use language (Iannàccaro and Dell'Aquilla 2001:267). The connection of language to place falls under the rubric of language ideology: people's beliefs about language use. The identification of features that distinguish any given group from others, whether the group is defined by ethnicity or geography, as is the issue in question here, or by gender or social class or some other social factor, relies as much on underlying beliefs about the people who use it as on linguistic reality. People tend to map their beliefs about each other—the boundaries of their communities—onto language. They tend not to be aware that they are doing this, however: generally, people believe that their evaluations of linguistic difference are objective and separate from their judgments of people themselves—language is, after all, a tool that we manipulate, not part of our genetic makeup. Therefore, judging someone's language is understood to be fair and unbiased. These facts allow us to ask about linguistic variation as a way of covertly asking about attitudes toward other people and about borders between different communities.

Over the years in Terrebonne-Lafourche, I have elicited information about ideology using exercises in perceptual linguistics. These exercises demonstrated that the strength of people's attachment to local French was clearly an attachment to the features that identify speakers as being from the bayou and not simply an attachment to ethnically defined communities that populate the region. Internal divisions, while important and meaningful, are secondary to place-based identity: there is a bayou identity shared by everyone.

WHAT IS PERCEPTION WORK, AND WHAT CAN IT TELL US?

Perception exercises provide valuable insights into the language ideologies that people may hold. We cannot get inside people's heads, but perception exercises can bring what are often subconsciously held beliefs about language into the open. Perception work encompasses a wide variety of exercises. Among them are exercises that ask people to describe the distribution of variation on the landscape (by drawing on maps or sorting place-names into piles, for example). Such exercises generally also ask people to describe in some way the speech varieties they are indicating via labels or examples. Other exercises, known as verbal guise or matched guise tests, involve the use of recordings, sometimes digitally manipulated, played back to listeners who then are asked to make judgments about the speakers (are they friendly? intelligent? reliable?) or guess at their demographic affiliation based on their speech alone (as I did with the test introduced in chapter 4). Still other exercises include asking people to rate dialects or areas in some way (good/bad, friendly/unfriendly, and so forth) on a Likert scale or asking people to imitate other dialects.

Preston (1986) points out that perception exercises help us to understand such concepts as the speech community. This is most broadly defined as people who share linguistic norms (Labov 1966, 1972), but whether members must speak alike as well to be included in a community is a matter of some dispute. Asking people to determine where people may speak differently, to evaluate speech as different or not, to express attitudes toward speakers based on their accents, and so on, may help provide a better understanding of whether similarity of speech is a crucial component. Or as Fridland (2008) puts it, such exercises can tell us which features are related to social embedding and used to organize our experience. In short, perceptual dialectology can tell us about the boundaries of communities: Who are the in-group members, and who are not? Which features do we use to identify them—or do we ignore variation? Perception exercises can tell us about the social associations people make with language and the attitudes they hold toward speakers.

Language ideology is often employed in disputes over identity and/or political autonomy: people may claim their language differs significantly from that of another group, while the other group simultaneously claims a lack of (important) differences. Preston (2011) cites in this regard Wolff's (1959) study of a Nigerian dispute in which one group wanted to assimilate another; the group wanting to assimilate claimed no language difference, whereas the group targeted for assimilation resisted on the grounds that because their language was entirely different, they could not be the same people. There are many other examples: in some cases, both groups involved in a dispute may claim they speak differently from each other. In one famous example, Czech and Slovak speakers, who once bristled at having to share a country, frequently claim that there are enough differences between their speech varieties to merit their designation as separate languages, which of course justifies their desire to live in different countries. More neutral residents of the former Czechoslovakia, however, frequently mock such a suggestion, arguing that while there may be a few differences between the two languages ("about three words," my Czech-raised father used to joke), Czech and Slovak are still simply dialects of the same language. In an example of the reverse effect, Zinsli (1957, cited in Sibata [1959] 1999) notes that several isoglosses pass through the city of Bern, Switzerland, but that residents do not acknowledge them; Irvine and Gal (2000) refer to this process of ignoring differences as *erasure*. Given that there is no mathematical formula for determining how much or which type of variation will result in the recognition of dialectal difference, perceptual exercises can be very useful in interpreting the results of research on linguistic variation.

VERBAL GUISE TEST

In the verbal guise exercise described in chapter 4, I asked people to identify speakers' social demographics. I conducted this test during my initial fieldwork as an add-on to the sociolinguistic interview. In lower Terrebonne-Lafourche, I asked participants to guess not only the speakers' ethnicity, as I had elsewhere in the state, but also their geographic origin—in which town had they been raised? While in part I was using the test to verify my findings (if I had found no significant variation but a high rate of accuracy in identifying people by their speech alone I would have revisited my list of targeted features), I was also using it to test the social significance of my results: There may be variation, but does it matter to people?

Much as was the case when I administered the test outside the lower Lafourche basin, the exercise was fraught with difficulty. In this case, it was because participants often knew the speakers and identified them by name as they were listening. One participant was related by blood or marriage to all but three of the speakers, and she knew one of the remaining three, who was a resident of her town. Even when people were not able to identify

Table 5.1 Perception Test Results, Ethnic Affiliation						
	Clip 1	Clip 3	Clip 4	Clip 5	Clip 6	Clip 8
Speaker's declared ethnicity (and origin)	Indian (Golden Meadow)	Indian (Pointe aux Chênes)	Cajun (Pointe aux Chênes)	Cajun (Dulac/Grand Caillou)	Indian (Île à Jean Charles)	Cajun (Golden Meadow)
% correct	61.3	52	77.4	71	70.8	79.3

the speaker, they were clearly trying to do so, making incorrect guesses at their identity and sometimes naming a relative of the speaker (a brother, for example) instead. Even when they did not immediately recognize the voice, when the exercise was over and I told them their results and gave them the speakers' names, participants frequently knew at least some of the speakers. Moreover, they were clearly basing their decisions on the content of the clips. Normally with such an endeavor, one would use the same text for all speakers. Given the fact of illiteracy in French, however, this option was not available, so I had selected clips in which people discussed everyday life in the area. Though I had taken great care to select clips that described activities I had witnessed or heard discussed widely, people clearly had ideas about where activities might be more common (up to and including such common activities as the keeping of vegetable gardens). One respondent, for example, decided that clip 6 must be from Dulac, because the speaker was talking about oyster fishing. In fact, that clip was from the Île à Jean Charles, and I met oyster fishermen on all four bayous.

Despite these difficulties (which led to my processing the data both with and without the results for speakers who were known to the listeners—I always eliminated those who were identified by listeners while the exercise was in progress), the results suggest that listeners are in fact reasonably good at identifying speakers' geographic origins and ethnic affiliation, though this statement is not without qualification. Table 5.1 shows listeners' attempts to guess at ethnic affiliation.

Two further clips complicated the situation. Speaker 2 refused to identify binarily and declared mixed ancestry, but listeners identified him as Indian at a fairly high rate. Speaker 7 presented himself to me as Indian, but the accuracy with which people identified him as such was very low. Table 5.2 shows the numbers for these speakers.

These numbers can be explained with some further background information, however. The speaker in Clip 7 was a man who was known to many of

Table 5.2 Perception Test Results, Ethnic Affiliation		
	Clip 2	Clip 7
Speaker's declared ethnicity and origin	Mixed	Indian
% Guessed Indian	69	36.7

my respondents as Cajun rather than Indian. Clip 2 features a man known as Indian to his neighbors. Their scores reflect these identities. Given that I had selected the clips with no idea as to which features might pattern with ethnicity, the results are in fact quite suggestive of people's ability to identify speakers' affiliations; most speakers are firmly identified by ethnic affiliation.

When asked to identify speakers by their town of origin, however, listeners were deeply unhappy, and the numbers that result show that they were essentially guessing at random. Only one speaker is correctly identified by town origin (a Cajun from Golden Meadow) at a rate that allows us to suggest it was not by chance. Consolidating the responses such that a guess of Dulac is coded simply as Terrebonne Parish results in much higher rates of accuracy, however, suggesting that while listeners may not be able to identify the specific town from which a speaker originates, they can identify the parish. Given that participants had a 75 percent chance of guessing correctly for all speakers from Terrebonne, however, it is not surprising that they did much better under those circumstances. While I suspect that redoing the exercise asking only for parish affiliation would result in much higher degrees of accuracy, I will not consider further their ability to identify speakers geographically here.

On the one hand, people were able to identify speakers' ethnicity with reasonable accuracy. On the other, a few participants simply labeled people they understood as "Cajun" if they identified as Cajun, and vice versa for self-designated Indian participants. Some overtly stated this as their rationale; one woman simply identified everyone as Indian (to the laughter and teasing of her family members observing) because she decided that they could be understood. At the same time, people who had asserted that there were no differences sometimes expressed surprise when I gave them their results, stating that "Those Indians sure sounded like Cajuns!" Statements such as this betrayed a belief that while there may be no significant differences between the groups, there *should* be. The question itself, however, seemed to prime participants to expect major differences—differences that they did not usually claim existed in the first place. The most common response when directly asked if differences existed was for participants to assert that there were no differences or that everyone spoke "about the same" (even when providing an example of a difference or two) and instead point to outside areas (Grand Isle, Lafayette, "up the bayou") as different. Further perceptual work, involving cultural domain analysis, helps to clear up this apparent contradiction.

CULTURAL DOMAIN TEST

During my initial fieldwork, I also conducted a more direct perceptual exercise in which I simply questioned participants regarding variation. The exercise was straightforward: I asked interviewees whether people "entour d'ici" (around here) all spoke the same way. I followed up when necessary with questions about specific towns and about Indians and Cajuns. Leaving the question open-ended in this way instead of providing participants with a list of places or a map to draw on allowed me to analyze the results using methods borrowed from cultural anthropology called *cultural domain analysis*.

What Is Cultural Domain Analysis?

Cultural domain analysis seeks to identify the categories that people from different cultures may use to create meaning. Generally (and best) applied to naturally occurring phenomena (plants and animals), it seeks to show the folk taxonomies that people use to semantically organize their world. So, for example, scientific taxonomy tells us that a cat belongs to the kingdom of Animalia, the phylum Chordata, the class Mammalia, the order Carnivora, the suborder Feliformia, the family Felidae, the subfamily felinae, the genus Felis, and the species *F. catus.* Very few people can rattle off this list of classifications, however; this is *not* how average people (or even biologists when not conducting research) conceptualize the world. If one asks a person on the street, they will respond that a cat is a kind of animal, in opposition to birds, dogs, and so on at the same level. A tree, however, is not an animal; it is a plant. Beyond this, tabby cats may differ from Persian cats, which may differ from Siamese, and so on (likewise, oak trees are different from birch, ash). These are the *folk categories* that people actively use in creating meaning, and cultural domain analysis seeks to access these categories.

In conducting this exercise, I essentially asked people to perform such a categorization activity. The goal in this case was to determine which items (people, in this case) fit into the categories of "sound like us" versus "don't sound like us." Because language is culturally imparted (i.e., learned and not innate), there is an assumption that if others speak like me, we must be the same; conversely, if others do not sound like me, they must belong to another group. The application of the methods of cultural domain analysis to phenomena that are not naturally occurring is imperfect; however, we can use it to a limited degree to help us analyze the results of perceptual exercises such as the one I conducted in Terrebonne-Lafourche. By asking who speaks differently, we are asking participants to show us the boundaries of their in-group (i.e. the folk category of "us") versus everyone else (i.e., "them") and any subgroups that may be contained in either category.

What Are the Methods of Cultural Domain Analysis, and How Are They Applied Here?

The two primary methods of cultural domain analysis are pilesorts and freel-ists. Pilesort activities ask participants to take a list of items and sort them into as many piles (categories) as they feel there should be and often also to name the categories if they can. One can repeat this exercise with a subset of the list until all items are split into piles that cannot be divided any further (which may mean each item is its own pile). Analysis of the resulting piles allows us to infer the reasons for the groupings at different levels. We can also compare the results across participants to see who clusters with whom. If we see a lot of agreement between participants, we can assume that the people interviewed belong to a single culture (or else that the target domain is universally organized the same way). If we see significant variation, we may be dealing with members of more than one culture. Freelisting simply asks respondents to list as many items belonging to a given category as they can without a list to aid the memory or to force a choice. Freelisting may also tell us about prototypical members of categories: an item that frequently appears on lists and/or tends to be listed early may be assumed to be more salient and therefore more prototypical of the category than those that are listed late or appear infrequently. In both cases, the researcher compares the results from several participants; as is the case with most things, individuals may show minor idiosyncratic variation in their evaluations. Many people may be willing to classify tigers and lions as cats, for example, while others may consider them to be different items on the same level (i.e., that cats, lions, tigers are all animals, as different from each other as they each are from dogs).

In performing the analysis I used Anthropac (Borgatti 1996), a program that helps researchers compare pilesorts and word lists from many par-ticipants. It shows which items may cluster together in opposition to other clusters at a given level of taxonomic classification as well as which people may cluster together based on their responses. In this regard, it can also check the results for consistency: if enough people give similar answers, the program determines that the participants likely belong to the same culture, since at least for the category under consideration, they organize their world in the same way. Conversely, if the answers are too disparate, the program suggests the possible existence of subcultures within the group under study.

My analysis has a few limitations. Several people did not answer the question, whether as a consequence of confusion or simply because the interview had to be ended early for one reason or another. Others answered it only in part (i.e., told me about ethnic variation but did not answer the question about geographic variation, or vice versa). A few answers had to be discarded for reasons detailed below. The exercise thus presented several difficulties in transforming the responses into workable data for the software. Still, I conducted a quantitative analysis that complements the qualitative discussion of results and provides some additional insight. A total of eighty-seven people answered the question in whole or in part.

Pilesort

I first treated the answers as though the participants had sorted a series of names I had provided into piles. This method necessarily excluded the question of ethnic variation, since it potentially exists on a different level: one may be embedded within the other—and indeed, two answers had to be entirely discarded because the only variation they noted was that "Indians in Golden Meadow [in the second case, Dulac] speak differently. Everyone else is the same," and I was unable to create a code for that type of answer that would be comparable to the rest of the data. Consequently, I excluded the ethnic data from the pilesort analysis and limited it to the freelisting. I began by creating a post hoc list of places: I listed all the locations mentioned and then merged a few that were variations on a theme. Thus I combined instances of "Lafayette" and "Gueydan" (both towns in western Louisiana) together under the generic "the West" (which was also given as an answer, with some frequency). I was also faced with a problem when dealing with "everywhere [in the parish/area/state] is the same" or "everywhere [in the parish/area/state] is different." When people gave such answers, I simply grouped all the towns named by other respondents into the same category (or not, depending on the answer), excluding any towns that respondents had specifically identified as unknowns. For example, the towns listed by participants in Terrebonne Parish were the Île à Jean Charles, Pointe aux Chênes, Dulac–Grand Caillou, Bayou Dularge, and Petit Caillou. When a respondent told me that all of Terrebonne Parish was the same, I listed all six towns in the same pile, unless follow-up questioning had led to something to the effect of "It's all the same, though I'm not sure about the Island," in which case the Island (or whichever town was actually named) was excluded. This system allowed me to treat the responses as though the participants had received a list with all the items on it and had simply chosen not to classify those they didn't know about. This approach is further justified by the fact that I generally asked more specifically about given towns after respondents had given their initial answers. It is an imperfect solution, but in that all answers were subject to the same coding, the results are systematic. Likewise, the data is potentially problematic in that the question asked whether people "around here" spoke differently, and many respondents chose to talk about France or Canada in their answer, while others did not, likely because (1) it is obvious that France and Canada are different, and (2) I asked specifically about local variation. However, when people did answer in this way, it was generally to show that local French was all the same—if I was interested in difference, I should look to the West or even to foreign countries. To deal with this problem, I ran the data a second time, this time excluding the answers about other places.

The pilesort results show unsurprisingly that the first distinction people make is at the national level: France and Canada (as a single group) are different from Louisiana. At the next level, France and Canada are differentiated from each other, and a few less-frequently mentioned places within Louisiana (Natchitoches, Houma, and St. Bernard and Plaquemines Parish) are distinguished from the rest of the state. When only areas within the lower bayous are included in the analysis, there is a clear Terrebonne versus Lafourche divide. Towns within Terrebonne Parish begin to be distinguished before those in Lafourche do, with Pointe aux Chênes and the Île à Jean Charles the last to separate (i.e., that they are the least different from each other and, conversely, the first to be grouped together by similarity). However, the local findings are colored by the fact that there were three times as many participants from Terrebonne as there were from Lafourche: residents of Terrebonne were more likely to treat Lafourche Parish as a monolithic entity, and vice versa for those from Lafourche. This was true for the freelist activity as well. Consequently, for both exercises I analyzed the area in the aggregate and then treated the results for each parish separately and for each town separately. For the freelisting activity I additionally analyzed each ethnic group separately and I compared ethnic groups across towns and compared them to each other within towns; for reasons I detail below, it was unnecessary to do this for the pilesort activity.

The most important finding of the pilesort was that there was very strong consensus among participants in Golden Meadow and Dulac–Grand Caillou respectively regarding geographic variation regardless of ethnic affiliation. The program declared that the results for those towns, when taken independently, "support the assertion that despite individual differences, all respondents in the sample belong to a single culture with respect to this domain." To be clear, this does *not* mean that there are no cultural or ethnic differences at all within Golden Meadow. It simply means that participants in the exercise are in strong agreement regarding regional variation. Participants in Pointe aux Chênes and the Île à Jean Charles (considered together, since the towns clustered together perceptually), while still fitting the consensus model, were less strongly in agreement. This makes intuitive sense, in that the Lafourche Parish line passes through Pointe aux Chênes, and both towns are situated at the approximate halfway point between Dulac and Golden Meadow. The feeling of distinctiveness would logically be somewhat attenuated by the presence of familial ties to both towns, and that the parish line passes straight through the middle of town may give rise to some confusion in answering—one participant noted, regarding the question of difference between Pointe aux Chênes and Lafourche Parish, that "on est proche tous parents [we're almost all related]." Perhaps most telling was the participant who, when asked whether Pointe aux Chênes was different linguistically from Dulac, responded that she had relatives there without commenting on their speech at all. Indeed, when all the towns are considered separately, Pointe aux Chênes is the only town that does not show at least weak consensus, and the participants do not cluster along ethnic lines in their disagreement. Many participants had relatives from Lafourche Parish in particular; this was far less common in Dulac–Grand Caillou, though there, as noted, many people had familial ties to Pointe aux Chênes and the Island, and residents were known across Terrebonne Parish and occasionally the entire region regardless. When all the towns were grouped together for analysis, however, the consensus model failed, and the program declared that I likely had subcultures within my sample. This finding is explained by the freelist results.

Freelist

When analyzing the data as freelist items, I treated them as answers to the open-ended question about differences as the category of "people or places that are different," which allowed geographic and ethnic variation to be treated at the same level. Participants listed Lafourche Parish, Terrebonne Parish, and various individual towns within the area (those under study and also others such as Petit Caillou in Terrebonne Parish, and Thibodaux, Larose, and Galliano in Lafourche) as well as ethnic differences. Some respondents suggested that every town had its own way of speaking, others that differences ran along bayou lines. Many respondents listed foreign countries (specifically France and Canada), and other regions or towns across the state, including Grand Isle; Plaquemines, St. Bernard, and Cameron Parishes; "the West"; Lafayette, Gueydan, Loreauville, and Breaux Bridge. As I did with the pilesort, I grouped references to prairie towns (Lafayette, Gueydan, etc.) with references to "the West" and I grouped the towns of Buras and Empire with Plaquemines Parish, in which they are found. Often, people simply listed places they had been to outside the area, saying they didn't know about other places, but they knew about these. I also listed "everywhere is different" and "everywhere is the same" (coded as "nowhere is different") as proper responses rather than divide up the towns as I did with the pilesort. With this regrouping I had a list of 30 total items; the average list was only 1.8 items long; the longest (only three lists) five.

I analyzed the results in the aggregate and then broke them down by town and ethnicity. Overall, the top five responses, in order of frequency, were as shown in table 5.3. For the region as a whole, the groups most frequently listed as different are defined by geography, not by ethnicity. That ethnicity appears before foreign countries on the list is unsurprising given the way the question was presented—I asked about people "around here," after all. As was the case with the pilesort, the results are somewhat skewed given the larger number of respondents from Terrebonne Parish. Table 5.4 shows the results

Table 5.3 Top Freelist Answers	
Lafourche	37.3%
The West	26.5 %
Cajuns/Indians	22.9%
France	18.1%
Canada	15.7%

Table 5.4 Top Freelist Results by Parish			
Terrebonne			
Item	**Frequency (%)**	**Average Rank**	**Salience**
Lafourche	54.4	1.19	0.516
France	22.8	1.62	0.172
Indians/Cajuns	21.1	2.58	0.096
West	19.3	1.55	0.16
Canada	17.5	1.7	0.127
Lafourche			
Item	**Frequency (%)**	**Average Rank**	**Salience**
West	42.3	1	0.423
Indians/Cajuns	26.9	2	0.16
Terrebonne	15.4	3.5	0.067
Nowhere	15.4	1	0.154
Canada	11.5	3.67	0.044

Table 5.5 Freelist Results Ranked by Salience			
Terrebonne			
Item	**Frequency (%)**	**Average Rank**	**Salience**
Lafourche	54.4	1.19	0.516
France	22.8	1.62	0.172
West	19.3	1.55	0.16
Canada	17.5	1.7	0.127
Indians/Cajuns	21.1	2.58	0.096
Lafourche			
Item	**Frequency (%)**	**Average Rank**	**Salience**
West	42.3	1	0.423
Indians/Cajuns	26.9	2	0.16
Nowhere	15.4	1	0.154
Everywhere	11.5	1	0.115
Grand Isle	7.7	1	0.077

broken down by parish. What emerges from this is that overall, residents of the lower bayous define themselves primarily in opposition to others geographically rather than by ethnicity. This is reinforced when the results are ranked in terms of salience, which takes into account the placement of an item on the list in addition to its frequency (table 5.5).

In both parishes, the salience of ethnic variation lags significantly behind regional variation. In Terrebonne, ethnic variation has now dropped from third place to fifth, behind additional regional variation, including both Lafourche Parish and the West. Of equal interest is that Terrebonne residents define themselves primarily in opposition to Lafourche. Meanwhile, residents of Lafourche primarily distinguish themselves from those in the West, with Terrebonne Parish appearing behind several other distinguishing factors. Though ethnicity is in second place in Lafourche, its salience lags well behind that of geography.

The results broken down by town and ethnicity (table 5.6) show the same repeated pattern. In every instance but one (Cajuns in Golden Meadow), residents list ethnicity below geography—in fact, in Dulac, ethnicity does not even appear on the lists created by Cajuns. These results, combined with the pilesort, corroborate cultural maps that suggest two poles for Louisiana francophone identity: "The West," often represented by the city of Lafayette, and lower Lafourche Parish ("the Bayou"). To residents of Terrebonne, it is important to distinguish themselves from the nearest major cultural hub. Residents of that hub are more concerned with contrasting themselves to the West than with noting that they are not from Terrebonne. They are next concerned with the fact that the Bayou has both Cajun and indigenous residents. Both of these groups comprise bayou identity. Terrebonne Parish, meanwhile, is seen by most as a part of the general area, worthy of distinguishing only at a microlocal level. Even for Cajuns in Golden Meadow (table 5.7), where ethnic differences tie the West for first place in frequency of response (37.5 percent), the salience of geographic difference is higher than it is for ethnic variation.

While these results show the general inclination toward regional oppositions taking precedence over ethnic oppositions, taken alone they tell a slightly misleading story. It is true that Terrebonne residents defined themselves in opposition to Lafourche, but the clearest statements of distance were reserved for the West and other locations outside the lower bayou. This was true for residents of both parishes, who described both intra- and interparish differences as negligible even when acknowledging their presence. The high rank of ethnic differences on the Lafourche Parish lists is also misleading: while differences were recognized, participants mentioned them

Table 5.6 Results by Ethnicity and Town			
Cajuns		Indians	
Golden Meadow	Frequency (%)	Golden Meadow	Frequency (%)
West	37.5	West	50
Indians/Cajuns	37.5	Nowhere	20
Terrebonne	25	Natchitoches	10
Everywhere	18.8	Everywhere Bayou	10
Canada	18.8	Indians/Cajuns	10
Pointe aux Chênes	Frequency (%)	Pointe aux Chênes	Frequency (%)
Lafourche	66.7	Lafourche	53.8
Indians/Cajuns	41.7	France	30.8
West	25	Indians/Cajuns	30.8
Canada	25	West	30.8
France	25	Canada	30.8
Dulac–Grand Caillou	Frequency (%)	Dulac–Grand Caillou	Frequency (%)
Lafourche	62.5	Lafourche	35.7
West	50	Every Bayou	28.6
Petit Caillou	37.5	France	28.6
Dularge	37.5	Indians/Cajuns	14.3
Morgan City	37.5	Canada	14.3
		Île à Jean Charles	Frequency (%)
		Lafourche	58.3
		Grand Isle	16.7
		France	16.7
		Every Bayou	16.7
		Indians/Cajuns	16.7

Table 5.7 Responses from Cajuns in Golden Meadow			
Item	Frequency (%)	Average Rank	Salience
West	37.5	1	0.375
Indians/Cajuns	37.5	2	0.229
Terrebonne	25	3.5	0.108
Everywhere	18.8	1	0.188
Canada	18.8	3.67	0.071

only following specific, targeted questioning that often asked if even minor differences existed between groups. For example,

(5.1)

R1: Et entour icitte tout le monde parle le même français ou euh	R1: *So, around here did everyone speak the same French or, uh*
. . .	. . .
L1: Oh ouais.	L1: *Oh yeah.*
R1: Et à Pointe au Chien aussi, euh	R1: *And in Pointe au Chien, too, uh . . .*
. . .	
L1: À Pointe au Chien aussi, ouais.	L1: *In Pointe au Chien too, yeah.*
R1: Et à Dulac.	R1: *And in Dulac.*
L1: Uh-huh, uh huh.	L1: *Uh-huh, uh-huh.*
R1: Et Grand Caillou.	R1: *And Grand Caillou.*
L1: Ah ouais.	L1: *Oh yeah.*
R1: Et est-ce que est-ce que y a des, des petites différences ou euh . . .	R1: *So, so were there small differences, or uh . . .*
L1: Euh, des petites différences, ouais.	L1: *Uh, small differences, yeah.* (Indian, Golden Meadow)

Literally nobody volunteered, unprompted, that Indians spoke differently than Cajuns, nor was it anyone's first answer when asked whether people spoke differently "around here." This was largely true for intraregional variation as well: Lafourche is the top answer for Terrebonne participants, and it sometimes was offered without prompting as a place attesting significant variation, but most often, as was the case with ethnic differences, it was dismissed as negligible. "Mais c'est proche tout pareil [But it's all about the same]," was the constant refrain. It was in fact difficult to classify a few answers, as people gave ambiguous answers and "C'est proche tout pareil" was offered both in support of an assertion of *lack* of difference and as a mitigation of an admission of difference. For example,

(5.2)

Y avait pas de différence. Le monde parlait proche un tout peu . . . tout le même français.	*There was no difference. Everyone spoke almost a little bit . . . all the same French.*

(5.3)

Les Cadiens? Et les Indiens. Um, proche que sûr bien presque pareil. Mais ouais.

Cajuns? And Indians. Um, I'm almost sure they're about the same. But yeah.

(5.4)

R1: Mais, euh, à Lafourche eusse parle pareil qu'ici ou euh . . .

L1: Ouais. Oh, peut-être des des choses un 'tit brin différent but c'est c'est le même français. C'est parler uh, le monde appelle ça parler cadien. *Cajun speak.*

R1: Mais y a quelques différences, tu dis, à Lafourche?

L1: Ouais un petit brin, c'est, uh, c'est joliment pareil.

R1: But, uh, in Lafourche do they talk the same as here or, uh . . .

L1: Yeah. Oh, maybe some some things are a little bit different, but it's, it's the same French. It's talking, um, people call it talking Cajun. Cajun speak.

R1: But there are some differences in Lafourche, you say?

L1: Yeah, a little, but it's really similar.

(5.5)

R1: Mais ici en bas, y a des différences, uh, dans le manière qu'eusse parle à Lafourche?

L1: Uh, pas plein, uh-uh, de la différence. C'est proche tout pareil. Tout les . . . comme ici, à Lafourche et 'Tit, 'Tit Caillou, Grand Caillou . . .

R1: Et Dulac, et . . .

L1: Ouais, et c'est proche pareil comme nous-autres on parle.

R1: Mmm-hmm. Tu peux connaître si quelqu'un devient de de Lafourche si tu les ente—attends parler, ou uh . . .

L1: Ouais. Uh, tu peux proche dire équand eusse parle si eusse vient de Lafourche ou, uh, Dulac, ou uh . . .

R1: But down here, are there differences, uh, in the way they speak in Lafourche?

L1: Uh, not much, uh-uh, difference. It's almost all the same. All the . . . like here, in Lafourche, and 'Tit, 'Tit Caillou, Grand Caillou . . .

R1: And Dulac, and . . .

L1: Yeah, and it's all about the same as we speak.

R1: Mmm-hmm. Can you tell if someone comes from Lafourche if you hear them speak?

L1: Yeah. You can almost tell when they talk if they come from Lafourche or Dulac, or uh . . .

(5.6)

R1: Mais, à Larose c'est différent?	R1: *So, in Larose it's different?*
L1: No, no eusse dit des mots différent.	L1: *No, no. They have some different words.*

In example 5.2, the speaker claims there are no differences and then follows up with "It's almost the same." In example 5.3, the speaker is *almost* sure they're *about* the same. The use of *proche* (almost) implies differences despite the claim that there are none. In example 5.4, the speaker initially says everything is the same, then backtracks and suggests that some things are different in Lafourche but mitigates this by noting that it is really similar. The speaker in example 5.5 confirms that everywhere is the same but then admits there are small differences and even suggests he can tell where someone is from by their speech. And in example 5.6, the speaker conflictingly claims that Larose (just up the bayou from Golden Meadow) is not different; they only say their words differently. The Anthropac software cannot account for the level of ambiguity or the circumstances of the description; excerpts such as these add nuance to the quantitative findings.

Even unhesitant claims of difference were qualified as being only "un 'tit brin différent [a little bit different]." Overall, participants were unwilling to give much importance to local variation. The hard-to-classify answers are not really ambiguous or contradictory, though superficially they may appear so and they were hard to enter into a program that did not allow for nuance. People are clearly aware of minor variation in the region but do not consider this variation to be important enough for the resulting speech to be designated "different" from their own. A few participants insisted in response to the question of difference that "C'est toujours le français [it's still French]," or "On est tous le même monde [we're all the same people]," despite the minor variation attested, with one going so far as to assert that "Les Indiens, oui, c'est tout, c'est tout Cadien ("The Indians, yes, they're all, they're all Cajuns]" (clearly not intended as a rejection of their indigeneity but rather a statement of inclusion in the in-group despite ethnic difference). This was true to a far lesser degree of the West or France. While a few participants felt that all of Louisiana was the same, most who cited the West either proposed it as a place to find *real* difference (or to find any at all in Louisiana) or noted that the French of the lower bayous was *not* the same as that found in Canada or France. In some cases, the interviewee who initiated the perception exercise by volunteering that French was different in the West or abroad. Alternately, I initiated the exercise by asking if there were differences "around here," the participant would say yes, and then when asked where, they would provide the West as their first example. One respondent who claimed that there were differences on every bayou, for instance, followed up by noting that the word for *turtle* should not be used in Lafayette (he meant *caouenne* [snapping turtle], which has a dual meaning on the prairie as a vulgar term for female anatomy). It was unclear at that point whether "every bayou" referred to the five major bayous of Terrebonne-Lafourche or if he was including the Teche, which flows through the prairie parishes around Lafayette, in his definition.

In fact, even when they concluded that variation existed within the bayou region, participants were rarely able to give concrete examples of it: only twenty-nine respondents gave any answer to the question, and five of them were "I don't know." Most are vague statements to the effect of "they use different words than we do" or "it's the way they pronounce their words." Those that provided concrete examples often offered words that were rude in some other part of the state or the world but not on the bayou (or vice versa) and that had clearly resulted at some point in hilarious miscommunication. Such obvious examples stick in people's memories but nonetheless were not examples of variation within the bayou region. Requests for examples of differences also resulted in responses citing nonlinguistic information, such as "We don't talk to people over there very often," "We weren't even allowed to mix with people from other towns" (which, it transpired, was really about ethnic groups mixing), or "Eux-autres s'appelle des Fourchons [They are called Fourchons]." These are all clearly examples not of attested variation but rather of patterns of interaction and the nature of relationships (often somewhat jocularly hostile; it is not necessarily the people of Lafourche who call themselves Fourchons). Others replied, when asked about a specific town, that they often visited or that they had family there, without answering the question about language; the implication is that they do speak the same.

CONCLUSION

The greater picture that emerges from this data is that there is a place-based identity, a bayou-based identity, that is shared by everyone. This is not to say that ethnic identity is not important—it is, very much so, as is evidenced by the ease with which participants were able to identify people's ethnicity by their speech. The Indian community's fight for recognition as indigenous is evidence alone. Further support comes from the names people give to their language. Everyone calls it French. To distinguish it from that spoken in

France, however, various other terms are used. These include Cajun French, *français plat* (flat French), patois, or even broken French (or sometimes, *baroque*) as a consequence of the English influence. Indian speakers were willing to accept the term *Cajun French*, with one going so far as to scoff at the notion of calling it Indian French simply because he, an Indian, would be using it. Most Indian residents, however, preferred to simply call it French, with no modifier (though references to *Indian French* have recently become more common), regardless of whether they recognized differences. The stated rationale was that they were not Cajun, so they did not call it Cajun French.

Many participants noted that they did not often get to talk with people from other places included in the study: Terrebonne residents said they rarely if ever met people from Lafourche, or people from Dulac commented that they never got to the Island or Pointe aux Chênes. Such comments show that lower-level distinctions are more often invoked on a day-to-day basis. However, when asked (by an outsider) where differences lie, they point first to another region, another *place*, and only later to the ethnic differences that are more often invoked. This evidence makes clear not that it is more important to people to be from the bayou than it is to be Indian or Cajun but rather that all people, whether Indian or Cajun, when asked to define themselves, first make clear their affiliation with place. It is important to be from somewhere, regardless of who one is otherwise. Communities are rooted in places. This affiliation with place is expressed when people describe the perpetual movement up the bayou to escape erosion and flooding and ultimately declare that this is just how life is not because they are Indians or Cajuns but because "On vit sus le bayou! [We live on the bayou!]." While lower Terrebonne-Lafourche is a clearly identifiable zone to outsiders and insiders, within it, the minor variation that distinguishes subdialects is salient to listeners but minimized ideologically. Everyone—Cajuns or Indian—is from the bayou first. This phenomenon recurs in the next chapter, which addresses the dispute over a town's name.

Chapter 6

OAK POINT OR DOG POINT?

The Importance of a Name

On one of my early visits to Terrebonne Parish, I had lunch at the home of one of my interviewees, a member of the Pointe-au-Chien Indian Tribe. Also at lunch that day was a Cajun friend of the family's from up the bayou. During a lull in conversation, I innocently decided to ask about the fact that the town we were sitting in seemed to have two names, something that had struck me as curious.

What followed was a heated debate, with one side (the Indians) adamantly arguing that the town was Pointe au Chien (Dog Point) and the other (the lone Cajun) insisting it was and had always been Pointe aux Chênes (Oak Point). There was also a lot of laughter, but as time went on it became clear that underneath it, they were all dead serious and that my question had inadvertently touched a nerve. Fortunately, to the best of my knowledge, no friendships were ended as a result of my questioning that day, but I continued to ask the about the town's name over the course of my fieldwork. What emerged was a picture of a long-standing dispute that seemed to pattern along ethnic lines. Ultimately, however, it was clear that this split was a by-product of ethnic divisions within a single community: both groups lay claim to the same physical space and use the same means, the same types of stories, to do so. But because place is so closely tied to personal identity, the competing goals of the two subgroups result in the need to characterize the place differently. The dispute over the name is a dispute not over boundaries or stewardship of the place but rather over its characterization. A single bayou identity still underlies other identities.

THE IMPORTANCE OF A NAME

Discussing a town's name may seem superficial or trivial, but place-names are common sources of contention. At a bare minimum, one may be easily marked as an outsider for failing to correctly pronounce a name. For example, research I conducted with students in New Orleans (Dajko et al. 2012) revealed that authenticity—being a *real* New Orleanian—was carried in part in knowledge of the correct (i.e., local) pronunciations of the city's name (there are several considered acceptable) and in those of street names within it. This is not unusual, though it may carry more weight in New Orleans than it does in many other places. It is also possible to go beyond simply marking outsider status and offending people with a mispronunciation. A classic example is that of the pronunciation of *Nevada* (the US state). While innocent mispronunciations are for the most part simply corrected, they are also used as fodder for political campaigns (one wouldn't want to vote for someone who knows so little about the state that they pronounce its name incorrectly; conversely, one cannot persuade one's constituents how much one cares about them if one cannot get something so fundamental right). In that context, such errors can also be the source of much irritation, as is made clear by letters sent to the *Reno Gazette-Journal* in 2003 regarding an article published—as one of the letters notes, on the front page of the paper—in response to a mispronunciation by then-president George W. Bush:

> True Nevadans find it unamusing and lacking in state knowledge when "they" pronounce Nevada with a long "a." . . . If one wants to be viewed as a Nevadan or ensure votes or business, they should be sure to pronounce the state name correctly. (Curtis 2003)

> It hurt me when I heard President Bush calling our state Ne-vah-da in the media. . . . We The People of the State of Nevada long ago decided how we want our name pronounced. We consider it a matter of respect when you live, work or visit here that you learn to say: "Nevada," not "Nevahda." (Van de Bogart 2003)

The issue came up again during the 2016 presidential campaign when Donald Trump used the incorrect pronunciation while boasting that he was using the correct one, ignoring the crowd's shouted objections. Once again the public was quick to condemn, in this case noting, in one online example, that "rookie mistakes liek [*sic*] this don't win you swing states" (Lordveus 2016).

Perhaps unsurprisingly, Trump did not win Nevada in 2016.

NAMING AND ITS ROLE IN PLACE-MAKING

If knowing how to pronounce the name of a place is important, acknowledging the correct name is even more so. Both are part of the same process of place-making: by giving it a name, one brings a place into existence. Basso (1996b) emphasizes the importance of naming alongside storytelling in creating place. As Christie (2009:349) puts it, "To say the name of a place, to tell a story about a place is to waken memory, conjure up everything that ever happened there, and make it present again to the community." Schreyer (2008:4, 6) notes that names can be "performatives of stewardship," which she defines as "actions that assert leadership and responsibility for a community's lands and resources and (re)define who has this responsibility in a particular territory." In short, people tell stories about and name spaces that are meaningful to them and in doing so create places. The act of naming a place makes the namers the owners or at least the stewards of the space—the people rightfully responsible for its care. Naming is consequently not only an act of identity but also a potential site for struggles over both identity and power. Because names both reflect and help shape the worldviews and/or values of those who coin them (Greider and Garkovich 1994:8; Basso 1996b; Christie 2009; Gruenewald 2003), the ability to name a place and have others acknowledge that name is empowering (Tuan 1991:685). This is why the renaming of places is not uncommon following sociopolitical revolutions. The perception is that by erasing symbolic references, links to the past may be severed (Tuan 1991:688): people with a new outlook have taken over and have relabeled (and in doing so redefined) their world to reflect their own values and experiences. Thus, St. Petersburg becomes Petrograd becomes Leningrad becomes St. Petersburg. In a particularly extensive case of renaming, following the French Revolution, the new republican government embarked on an ambitious place remaking campaign, renaming many of the streets of Paris and erasing visual symbols of the monarchy such as the fleur-de-lis (a symbol that remains contentious to this day). Their efforts extended even to the old limestone quarries that underlie the modern city, twenty meters below the surface (a place frequented by few, even before the mid-twentieth century, when fines were imposed for entering), where, though the tunnels do preserve some of the prerevolutionary street names, only ten engraved fleur-de-lis remain across the vast network, all saved because they were in some way obscured from view at the time of the purge (Thomas 2018).

It may take a sociopolitical revolution to rename a city or nation (or at least the streets of a capital city) (Tuan 1991:668), but Hendry's (2006) study of the Riojan winemaking region demonstrates that the same power and identity struggles that result in large-scale renaming work on smaller scales as well. Those who name a place are its legitimate occupiers; to deny someone the right to name is to deny them a place. The evaluations of a speaker's authenticity or offense taken at a perceived lack of respect that the pronunciation of a name may incur highlight the fact that place-making is a deeply personal act; Low (1994:66) in particular stresses the "embeddedness of person, space, and action" in the creation of place.

To summarize, our experiences, our relationships with other people, the cultural practices that we learn as children (or sometimes as adults when we move), and the important events of our lives are all set in a space, and that space takes on meaning and becomes a place as it becomes intertwined with the events that make us who we are. To *name* a place is to claim stewardship over the space in which one's identity is anchored or to at least to make a claim to a connection to that space, and the recognition of one's right to name a place is empowering. If one's name is not recognized, one's right to that space and consequently one's very identity may be threatened. In lower Terrebonne Parish, such a struggle has been playing out for more than a hundred years.

THE HISTORY OF THE NAME OF POINTE AU(X) CH(I)EN(ES)

It is hard to argue that one or the other of the two names for Pointe au(x) Ch(i)en(es) is the obvious original; it is equally hard to establish just when or how the dispute over the name began. Documentary evidence exists for both names going back at least 150 years—essentially to the founding of the town. Though the town itself rarely appears on maps because it is considered officially a part of neighboring Montegut, maps—even modern ones, including Google Maps and the satellite map I bought at a marina in Terrebonne Parish—often list the body of water the town sits on as *Bayou Pointe au Chien*. The US Board on Geographic Names (GNIS), the federal body that maintains a database of feature (town, river, mountain, even building) names with the goal of promoting standardization, lists not only current names but also modern and historically attested variants for them. In this case, the database contains nine different versions of the bayou's name, including *Bayou Chene*,[1] *Bayou Chien*, *Bayou Chin*, *Bayou du Chien*, and variations on *Bayou Point(e) au(x) Chien*. Fully three of these (*Bayou Chene*, *Bayou Chin*, and *Bayou Point au Chien*) are attributed to a single source: the 1963 edition of

Read's *Louisiana-French* [(1931) 1963]. Read himself gives the earliest citation as *Bayou Chin*, attributed to an 1831 land survey. By 1856, he notes, it is *Bayou Point(e) au Chien* (Read [1931] 1963:178). The earliest map I have been able to locate that names the bayou is Bayley's (1853) map of Louisiana, on which it is *B. Chene*, as it is on Johnson's 1866 map. Only four years later, Hardee's 1870 map of New Orleans and the surrounding area labels it *Bayou du Chien*, as do Hardee's 1871 map of Mississippi and the National Map Company's 1927 map of Louisiana. Rand McNally's 1889 map labels it *B. Pointe au Chien*, a 1909 map (Rootsweb n.d.) has *Bayou Chêne*, Bocage's 1916 map shows *Point-au-chene*, a 1930 National Geographic Society map has *B. P. aux Chenes*, and Rand McNally's 1939 map lists *Bayou Chien*. Nearby is a Lac Chien (whose name has gone uncontested), lending some credence to the *Chien* version of the story, though the bayou does not meet the lake in any way; they are merely close to each other.

Legal documents offer a little more insight into the time depth of the disagreement. An 1882 Louisiana Supreme Court Case intended to judicially settle a dispute over the location of the eastern boundary of Terrebonne Parish notes that the bayou, which is otherwise unnamed, is "sometimes called Bayou du Chêne" (*Parish of Lafourche v. Parish of Terrebonne* 1882:3), as does an 1896 case (*Police Jury of Parish of Lafourche v. Police Jury of Parish of Terrebonne* 1896). An 1897 case, however, offers variants, noting that "on the southwest of Lafourche and the southeast of Terrebonne was a stream called 'Bayou du Chene' (otherwise known as 'Bayou Point au Chien,' and sometimes as 'Bayou du Chien')" (*Police Jury of Lafourche v. Police Jury of Terrebonne* 1897:2). Meanwhile, an 1892 case simply calls the bayou *Point au Chien* with no alternative name given (*Calder v. Police Jury* 1892). This last one is the second-earliest example I have been able to find of the use of *Point(e)* in any form (the earliest is the map from two years earlier).[2] The next instance of that term occurs in Swanton's field notes from 1907, where it is *Point au Chien*.

It is also hard to argue that there is an easy linguistic answer to the question. Which is a more likely progression: the addition of a glide and the deletion of word-final /n/ (chênes [ʃɛn] > chien [ʃjɛ̃]), or the deletion of a glide and the addition of word-final /n/ (chien > chênes)? (The vowel that precedes the /n/ in Louisiana French would be nasalized in anticipation of the nasal that follows, thus eliminating [de]nasalization as a step in the process, and the orthographic *-s* on *Chênes* and *-x* on *aux* are silent). That *Pointe au Chien* is in the singular (versus the plural *Pointe aux Chênes*) is easily attributable to either anglophone mapmakers and/or ambiguity in French: *Pointe aux*

Chiens would be pronounced identically. Historic examples give both *Chien* and *Chêne* in the singular, with a clear singular determiner to eliminate any confusion. It is plausible that the split came from a mispronunciation at some point in time, but in that case the change would have resulted in the transformation of one common word to another with drastically different meaning in the native language of the speakers (for whom it should therefore have been immediately apparent). For his part, Read ([1931] 1963:179), is certain the original name was *Chien* and attributes the confusion to carelessness on the part of the surveyors. Other plausible scenarios also present themselves, however. One is that an anglophone mapmaker with some minimal knowledge of French heard and therefore wrote *Chien*, given a greater degree of familiarity with that word than with the species-specific *chêne* and that this version subsequently was seen as official and adopted by francophones who had until then said *Chêne(s)*. A research report submitted to the Louisiana Department of Public Works in 1944, in discussing the history of the naming of the bayou, notes somewhat confusingly that it was called *Pointe au Chien* in honor of the oak trees that grew there, which the author translates as "'bois des chien,' in French" (Lombard 1944:6). His translation of *tree* as "bois" rather than the standard "arbre" would suggest that he was at least familiar with Louisiana French, possibly as a result of interviewing locals for his report. Whether he was a semiliterate francophone or an anglophone with limited ability in French, however, I cannot say; in any case, the anecdote supports a scenario involving literacy or lack of fluency in French as the likely root of the problem. Another similar possibility, of course, is that American Indians using French as a second language picked up the name as *Chien* rather than *Chênes*; conversely, the Indians who named it might have pronounced it in a way that sounded ambiguous to incoming whites and native French speakers, who selected the more bucolic *Chênes* in response, and that this is how the variant spread.

All this is speculative, however; in any case, it is clear from the documentary evidence that confusion over the name dates to at least the 1830s. It is also clear from the GNIS data as well as from modern pronunciations (nearly everyone I interviewed spontaneously used *Chien*) that *Chien* was the most common—though not exclusive—pronunciation by the mid-twentieth century. Nevertheless, the debate was "officially" cleared up in favor of *Chênes* in 1978, when local resident Laïse Ledet (who passed away a few years before I began work in the area) convinced the Terrebonne Parish Police Jury to have the town formally designated *Pointe aux Chênes*; the road signs were changed eight years later (Associated Press 1986).[3] The names on the water tower, school, and local fire station had already been changed, also through Ledet's efforts, at the time of the vote (Associated Press 1986), which was the culmination of a decades-long campaign to establish *Chênes* as the official name (though it is generally characterized as a name change). The campaign was so important to her that it is the focus of her obituary, the title of which is (cleverly enough) "Pointe aux Chenes Wouldn't Be the Same without Laise Ledet": it notes that she "fought vigorously to ensure the community's name was spelled and pronounced correctly" (Thurston 2002). In 2001, the Louisiana legislature approved a motion to officially designate the Wildlife Management Area that surrounds the town *Pointe aux Chênes* in accordance with the names on the other public edifices and following "the will of the people" (Dupre 2001:1). Today, large letters on the water tower still proclaim the town to be *Pointe aux Chenes*, reminding everyone that while many—if not most—residents continue to say *Chien* both in French and in English,[4] at least according to the official record, they are wrong.

INSIGHTS FROM THE LINGUISTIC LANDSCAPE

That the legislative action did little to settle the issue in the minds of many residents is clear from the argument I witnessed at the lunch table and from comments like the footnote found at Rootsweb's index to the online PDF copies of Ledet's (1982) genealogy, *They Came, They Stayed*, which observes that "the name of the area is a bone of contention to a number of people" (Hicks n.d.). An examination of the linguistic landscape (Landry and Bourhis 1997), the writing in public spaces on the bayou, will help to clarify the lines along which the argument is largely drawn.

Defining the Linguistic Landscape

The examination of the role public writing plays in the environment has been gaining in importance since the early 1990s, when scholars studying multilingual societies became interested in the way that different languages' occurrence in visible public spaces indicated their relative presence, prestige, and power in the community. Landry and Bourhis (1997:25) define the linguistic landscape as comprising "the language of public road signs, advertising billboards, street names, place names, commercial shop signs, and public signs on government buildings." Other researchers have employed broader definitions, including in their descriptions graffiti (Calvet 1990), "unfixed" or "mobile" texts such as bank notes, product labels, pamphlets, stamps, tickets, handbills, and flyers (Sebba 2010:59), and even nongraphic texts such as "verbal texts, images, objects, placement in time and space as well as human beings" (Shohamy and Waksman 2009:314). Per Landry and Bourhis (1997), signs posted by private citizens present a more accurate representation of linguistic reality than do those of government officials, who may wish to promote (or not) the importance of a language. While the study of the linguistic landscape was conceived to examine the role of different languages in multilingual communities, it is also a useful framework for considering the importance of variation—and in this case a single variable—as well. The appearance of a place-name on the landscape can tell us about the presence of a variant and about who is using a variant. In my examination of the naming of the town of Pointe au(x) Ch(i)en(es), I limit myself to instances of the name of the town on maps (including online maps) and on permanent edifices, both those placed by governmental agents (road signs, the water tower) and private citizens (supermarkets, homemade signs).

The Linguistic Landscape of Pointe au(x) Ch(i)en(es)

Though the water tower, fire station, school, and official road signs advertise the town name as *Pointe aux Chênes* and the bill requesting that the Wildlife Management Area be redesignated accordingly notes that the official state map published by the Louisiana Department of Transportation also lists the bayou by that name, one can still find maps showing the alternate name. When this occurs, one often finds both names juxtaposed on a single page or screen. Google Maps (probably the map service used most by visitors to the area today), for example, gives the name of Highway 665 as it passes through upper Pointe aux Chênes as *Pointe aux Chenes Road* but labels the bayou *Pointe au Chien*, with the two names listed adjacently, sometimes many times over, resulting in a somewhat comical visual reproduction of an old argument (figure 6.1).

Meanwhile, Pointe aux Chenes Road (labeled *Lower Hwy 665* after it passes through the open marsh and generally referred to as such by residents) passes by the Pointe au Chein Reserve boat launch before ending at Pointe au Chein Landing, which boasts the Pointe aux Chene Marina. These were, of course, named not by officials but by private citizens; the marina also maintains a Facebook page with that spelling. When driving along the road, one passes by the water tower as well as buildings labeled *Pointe aux Chenes Elementary, Point au Chene Volunteer Fire Department,* and *Pointe*

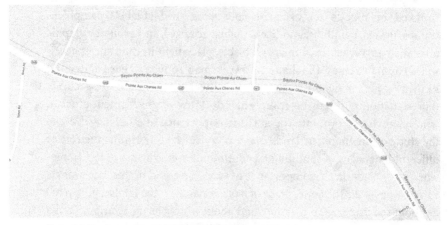

Figure 6.1. Google map showing both names. Google Maps 2017.

aux Chenes Supermarket, all along the upper portion of the bayou. Until 2008 one also passed, at the top of Highway 665, a small handmade sign, about the size of a real estate yard sign, welcoming motorists to Pointe aux Chênes and bidding them farewell on the opposite side (figure 6.2).

After passing through the open marsh, the entry to the settled area of lower Pointe au(x) Ch(i)en(es) is marked by a large sign welcoming visitors and residents to the home of the Pointe-au-Chien Indian Tribe. The sign was designed and commissioned by the tribe and placed right in front of the bridge leading across the bayou to Oak Point Road in 2008. On Google Maps one also finds a reference to the Pointe au Chien Community Center along the lower portion of Highway 665.

In short, the official name (though spelled in various ways, it is clear that *Chênes* is intended) features prominently on road signs leading to the town, the water tower, the school, the grocery store, signs made by the Wildlife Management Area, the fire station and marina, and, the small handmade sign (until 2008). It is clear to anyone who enters the town that the *Chênes* variant has official backing, and the visual evidence seems to indicate that *Chênes* is the more common pronunciation.

The *Chein* spelling that occasionally appears (and historically *Chin* as well) is indicative of a general confusion among people who do not speak French and/or are ambivalent about the town's name. It is unclear whether it is an attempt to spell *Chien* in which the letters are transposed, whether people did not know how to spell *Chênes* (less likely given the numerous examples, despite the slight misspellings), or whether the writers were aware of two spellings, were confused, and found a middle ground. In any case, it

Figure 6.2. Old welcome sign at top of Highway 665. Photo by the author.

Figure 6.3. Pointe-au-Chien sign (with researchers, for scale). Photo by the author.

is a visual reflection—intentional or not—of the ambiguous pronunciations some residents prefer to avoid taking a stand.

The welcome sign at Oak Point Road (figure 6.3) is the most important act of visual claim-staking on the bayou in recent years, on par with Ledet's efforts to get the school board and the water tower on board with her version of the name early in her campaign. The sign was an item on the agenda at Pointe au Chien Indian Tribe meetings for months before it finally appeared on the bayouside; the design and the wording were important topics of discussion that held up the sign's purchase for some time. The need for a sign was always stressed when the topic came up; one does not need to be a linguistic researcher to recognize the legitimacy granted by public signage.

The sign, which reads *Pointe-au-Chien Indian Tribe Community*, serves multiple purposes. It simultaneously welcomes visitors to the tribe's lands and makes a claim to independence from the more widely known United Houma Nation. Most important, however, it loudly proclaims the name of the town (via the name of the tribe) to specifically be *Pointe-au-Chien*. (It also proclaims French to be the town's language in that it welcomes visitors in French.) The sign is large: approximately six feet by three feet. In addition to the writing, it features a photo of a few trees set along a body of water, a dog, and a pirogue, with the corners marked by a shrimp, an oyster, a fish, and a crab—sources of income both historically and today.[5] This sign contrasts sharply with the small sign that marked the entry and exit to Pointe aux Chênes until 2008, when it fell apart due to age and weathering. That sign was discreetly placed on the opposite side of the ditch on the right-hand side of the road (when entering the town), at the top of Highway 665 where it branches from Highway 55 at Bayou Terrebonne. One side of this homemade painted sign featured a picture of a swamp with a boat to which was attached a fishing line in midflight with a tarpon (fish) attached; the other side depicted some reeds with a wading bird and the words *Y'all come back*. This sign served much the same purpose—to announce the name of the town—but its imagery was focused more on leisure activities[6] and on the beauty of the natural landscape; both the image on the front and the parting words on the back are aimed more at visitors than locals. In both cases, however, the natural resource of the area, the land itself, is the focus of the imagery, and the act of claiming the land—whether for Indians or not—is tied up in the name of the town.

The act of placing a large sign, which functions to rival the large official signs sporting the alternate spelling, at the entry to the lower settled area, functions to claim the space for Indians: it is named by Indians and therefore created by them. Thus, they are its stewards. This is further supported by the incorporation of storytelling including the explanation of how the town got its name in the first annual Pointe-au-Chien Indian Tribe culture camp in 2012 (Pointe au Chien Indian Tribe 2012). The story was included along with other stories from elders that described education and growing up in palmetto huts and was presented alongside activities like drumming, singing, and traditional crafts. It is in fact the first activity named in the list of activities in the press release describing the camp. Shana Walton (2017a), conducting research on subsistence agriculture in the area, reports that residents have told her that *Chien* is "more Indian" and that those who support *Chênes* are insecure in their identity or trying to prove their Cajunness. They are also linguistically insecure, insisting on correcting the French of others.

COMPLICATING FACTORS FROM INTERVIEW DATA

While the linguistic landscape suggests a simple Indian/Cajun divide with an overall preference for the *Chênes* variant, my recorded interviews reveal the complexity of the situation. An analysis of interviewees' stories regarding the town name's permutations shows how connections between identity, land, and language are made in Pointe aux Chênes and may shed light on why the confusion over the name has persisted for so long. *All* residents of the bayou, whether Indian or Cajun, claim a connection to the land, via their convictions regarding the name of the town. Unlike in other places, however (e.g., Robeson County, North Carolina, as presented in Blu 1996), there is no apparent difference in the way that ethnic groups construct place via language.

In asking which version was correct, I naturally found a wide array of opinions ranging from complete lack of understanding of the issue (the interviewee didn't seem to understand that there were people who might pronounce the name of the town differently, much less have an opinion as to which was right) to relative indifference (though how much of that results from the dual effects of being audiotaped and of speaking to an outsider is unclear) to strong support of one variant or another. The recordings document the nuances of the simple observation that the dispute seems to run along ethnic lines: both Cajuns and Indians argue in favor of *Chênes*, sometimes in strong terms, sometimes with less commitment. Both Cajuns and Indians also accept *Chien* with varying degrees of conviction. In fact, the strongest pro-*Chênes* response was from an Indian resident, as was the most strongly pro-*Chien* position. There are a roughly equal number of ambivalent responses from both groups. Opinions did not necessarily match production, for that matter. Out of thirty-nine natives of the town interviewed, twenty-two spontaneously pronounced the name during their interviews (the number of people who never said it unprompted is so high as a consequence of the shortened French version *La Pointe* [The Point], which neatly avoids the issue). Seventeen of the twenty-two (77 percent) said *Chien*, but seven of those seventeen went on to argue, when asked, that *Chênes* is correct, though they admit to using the other term. Only two people used *Chênes* spontaneously, and one additional person only used the full name of the town, pronouncing it *Chênes*, when I asked about the alternate terms; he never produced it unprompted. Two people refused to commit, opting instead for an ambiguous *Chiennes* [ʃjɛn], a pronunciation that may acknowledge both versions simultaneously.

Twelve people answered questions about the town's name when asked, with twice as many defending *Chênes* as supporting *Chien*. Such low numbers make it hard to suggest that this is meaningful, however. The number of people who were ambivalent about the issue equaled the number defending *Chênes*. Really, though, the responses can be ranked on a scale that progresses from a complete lack of expressed interest to possible resentment couched in diplomacy. For example,

(6.1)

> L1: The schoolteacher, Miss Laïse Ledet, you—you maybe knew her or heard of her, and she wrote a book, but anyhow she's the one that want to change it—change—she said that wasn't correct, that wasn't, it, you know it was Oak. Pointe of the Oaks you know so . . .
>
> R1: Hmm
>
> L1: She fought for it, and she finally had it done, and they changed the name. So. I don't know, so we have to get used to saying that, but a lot of times we still say the same old thing.

This response questions the validity of Ledet's claim. The speaker reports that "she *said* that wasn't correct," as opposed to "she learned" or some other verb that would imply agreement on the part of the speaker. Though the speaker's tone suggested that term didn't much matter to her, and she had previously made several light jokes about people liking dogs, her attitude toward it is betrayed by "so *we have to get used to* saying that." The sense of obligation implies a level of resentment. The "I don't know" that precedes it supports this: the expression serves as a placeholder while the speaker searches for words, implying an attempt to be diplomatic in doing so. I classified this response as ambivalent, but it was really a borderline pro-*Chien* answer.

The most solidly pro-*Chien* position was articulated by Vincent, an Indian shrimp fisherman who began his interview—which was videotaped—by announcing that he was from "La Pointe" and adding after a pause, unprompted, "La Pointe au *Chien*," heavily stressing the *Chien*. An hour later, the subject was brought up deliberately, and we had the following exchange:

(6.2)

R1: Et icitte le: le village ici, c'est Pointe aux Chênes.	R1: And this uh, this town is Pointe aux Chênes.
L1: {face drops} Pointe au Chien.	L1: {face drops} Pointe au Chien.
R1: Pointe au Chien?	R1: Pointe au Chien?
L1: {stony-faced} Chien.	L1: {stony faced} Chien.

R1: Cofaire eusse dit Chênes?

L1: C'est une vieille femme qu'a changé ça, mais it'*ll always be, uh, Dog Point*. La raison c'était la Pointe au Chien, uh (longtemps) XX pour la Pointe au Chien premier, premier temps que le monde a menu, XX (tu connais) comme un XXX qui restait dans le bayou étou (et tout?) là? C'était juste des plus des indi—juste des indiens ici proche.

R1: Uh-huh

L1: Et tout qu'eusse faisait était chasser. Tu vois, eusse avait, XXX famille des maisons ici x cinqs enfants peut-être ou six enfants. Tous les enfants avait des chiens, le monde avait des chiens. Et (puis icitte) x y avait plus de chiens que des, du monde.

R1: Mmm-hmm.

L1: Eusse avait des chiens chaouis, des chiens XX, des chiens chevreuils. Et là les enfants, les plus vieux petits enfants avait des chiens. Y a un tas de chiens. Eu—eusse appelait ça *Dog Point*, mais eusse appelait ça Pointe—la Pointe au Chien. {taps foot twice} {shaking head} *not Oak*

R1: Uh-huh?

L1: {shakes head}

R1: Et cofaire asteur c'est Chênes?

L1: Une vieille femme qu'a changé ça. La vieille femme, XX Laïse

R1: *Why do they say Chênes?*

L1: *It's an old woman who changed it, but it'll always be, uh, Dog Point. The reason why it was Pointe au Chien, uh, a long time ago when Pointe au Chien was first, when people first came here, XX (you know) like a XXX who lived on the bayou and all? It was just Indians around here.*

R1: *Uh-huh*

L1: *And all they did was hunt. You see, they had, XXX family had houses here x five, maybe six kids. All the kids had dogs, people had dogs. And (so here) x there were more dogs than people.*

R1: *Mmm-hmm.*

L1: *They had raccoon dogs, they had XX dogs, they had deer dogs. And so the kids, the oldest kids had dogs. There are {presumably were} a lot of dogs. They called it Dog Point, they called this Pointe—la Pointe au Chien // {taps foot twice} {shaking head} not Oak*

R1: *Uh-huh?*

L1: *{shakes head}*

R1: *So why is it Chênes now?*

L1: *An old woman changed it. The old woman, XX Laïse Ledet? It*

Ledet? XX Laïse Ledet qu'a changé ça, ouais.

R1: Hmm.

L1: Connais pas (quand ils ont) changé ça mais, *Lake Chien* là aussi Lac Chien, c'est tout Lac Chien. Eusse a pas changé Lac Chien, non.

R1: Ah, ouais?

L1: Mais eusse a mis ça ici *Oak Point*. Connais pas cofaire eusse a fait ça

L2: Et y a aussite Bayou Chien.

L1: Ouais. Mmm-hmm

R1: Àyoù eusse est asteur, les chiens?

L1: Eh là. Oh, le monde avait x les chiens chasse plus.

R1: Ah, ouais. Les chiens sont tous mort?

L1: Sont tous mort {cracks a smile, probably in response to everyone else's stifled laughter} Mmm-hmm. Sont tous mort.

was Laïse Ledet who changed it, yeah.

R1: *Hmm.*

L1: *I don't know when they changed it but, Lake Chien there's also Lake Chien. It's Lake Chien. They didn't change Lake Chien, no.*

R1: *Oh, yeah?*

L1: *But they called this Oak Point. I don't know why they did that.*

L2: *And there's also Bayou Chien.*

L1: *Yeah. Mmm-hmm*

R1: *Where are the dogs now?*

L1: *Now, that. Oh, people had x dogs but they don't hunt anymore.*

R1: *Ah, okay. The dogs are all dead?*

L1: *They're all dead {cracks a smile, probably in response to everyone else's stifled laughter} Mmm-hmm. They're all dead.*

It is immediately clear that the name of the town is very important to Vincent. He stresses it even before I and my two co-interviewers can pronounce it one way or another. When I first ask about the town name, his face falls as he answers, and he fixes me with a heavy stare (almost a glare) for the next several seconds as I repeat the name as he pronounced it and he confirms my pronunciation. A few minutes later, he makes sure the point is clear by repeating the name *Dog Point* twice in English (where there can be no misunderstanding given the drastic difference in the form of the words) and then adding, again in English, "*not* Oak."

He tells an elaborate story attributing the name to the historic presence of dogs. Both sides have backstories to explain their name of choice. The *Chênes* side, of course, posits instead the historic presence of oak trees (presumably live oaks). The story appears in Lombard (1944) and again in my interviews:

(6.3)

L1: It's supposed to be Chênes you see w—way down?

R1: Uh-huh.

L1: That's all they had, it was oak trees. All the way until, XX way past
the lake, you know, there, there used to be oak tree all the way.

(6.4)

Il avait un tas des chênes, des bois, des chênes à la Pointe? C'est pour ça que c'était nommé la Pointe aux Chênes.	*There were a lot of oaks, of trees, of oak trees on the Point? That's why it was called Pointe aux Chênes*

(6.5)

Point au—oui—on disait Pointe au Chien. *But* c'est c'est *really* le Pointe aux Chênes *because* y avait plein des chênes. Je pense c'est pour ça c—c'était mais qu— au commencement de: m—uh x *a long time ago the people would say* Pointe au Chien. But . . . c'est *really, uh,* Pointe aux Chênes *because o—of all the oak tree.*	*Point au—yes—we said Pointe au Chien. But it's it's really Pointe aux Chênes because there were lots of oaks. I think that's why it was so, but in the beginning of— uh, **a long time ago** the people would say Pointe au Chien. But . . . it's really uh Pointe aux Chênes because o—of all the oak tree.*

The stories told here are akin to the myths of place-making described in Basso (1996b) and Hendry (2006). Vincent not only bases his defense of the *Chien* variant in an appeal to a sort of creation myth but in doing so specifically invokes the ancestors: he asserts clearly that Indians were the first people to live here and that they named it. In making this assertion, he is claiming the space for himself as an Indian and a descendant of the first people to claim stewardship over the place. To rename the town is therefore to rob it of its Indianness. By extension, as an Indian, his own connection to this place is also unacknowledged and invalidated if the town does not bear the name that he knows and that his people gave it.

This appeal to the ancestors is manifest in the solidly pro-Chênes argu-ments as well. Laïse Ledet told the *Baton Rouge Advocate* in 1986, following the replacement of the highway signs, "We know that *our ancestors* named this community for its beautiful oaks that were lining the banks of the little bayou that bisects the ridge, or 'pointe'" (Associated Press 1986; emphasis added). In my own recordings, interviewees supporting *Chênes* likewise appeal to ancestry:

(6.6)

R1: Et **ta mère** disait la Pointe aux Chênes.	*R1: And **your mother** said Pointe aux Chênes.*
L1: Pointe aux Chênes.	*L1: Pointe aux Chênes.*

While Ledet's claim is to "our ancestors" the speaker in (6.6) makes a more immediate claim, insisting that her mother said *Chênes*; the implication here is that not only her mother but also all those before her also said *Chênes*. The appeal to ancestry echoes the situation described in Schreyer (2008:20), which explains that the Taku River Tlingit rely on ancestry in making their claim to being the rightful namers of the place: "The names that our people gave certain places, they gave that name because it had a *meaning*." For both camps on the bayou, the right to name the town is derived through inheritance, and each has a story that demonstrates that the name had meaning for those who originally named it—their own ancestors. More than just individual identity is at stake; an entire community's claim to the place is contingent on the acceptance of the right name.

That said, my interviewees attributed the name change entirely to Ledet rather than to a large movement, though Reggie Dupre's 2001 legislation requesting the change to the name of the Wildlife Management Area mentions following "the will of the people" (1) and Shana Walton (2017a) reports a strong pro-Chênes presence in Bourg (just up the bayou from Pointe aux Chênes) in the early 1990s, fifteen years before my research began. The attribution to Ledet occurs in discussions with people across the spectrum. Several interviewees make the observation, with one essentially suggesting that she alone found the *Chien* variant at all problematic:

(6.7)

Pointe au Chien, *yeah. They*— c'était, on disait ça vite. Et on—nous-autres, on pensait rien de ça mais elle, alle trouvait que c'était une insulte.	*Pointe au Chien, yeah. They—it was, we said it fast. And we—us, we thought nothing of it, but she, she found it insulting.*

(6.8)

> L1: She's the one that want to change it—change, she said that wasn't
> correct, that wasn't it, you know, it was Oak—Point of the Oaks, you
> know, so . . .
>
> R1: Hmm
>
> L1: And she fought for it, and she finally had it done, and they changed
> the name.

The interviewee in (6.7) did admit, after further questioning, however, that
Ledet was not alone in this opinion.

SEMANTIC FACTORS

One final factor must be addressed to draw any conclusions from the situation: the potential for humiliation given the implications of the town's name.
The notion that dogs are degrading in some unspecified way frequently underlies the pro-*Chênes* arguments and is ultimately very telling. An example
comes from the text of the 2001 bill requesting that the Wildlife Management Area be renamed. The fourth bulleted point simply states, "WHEREAS,
Pointe-aux-Chenes' English translation means Point of the Oaks and the
English translation of Pointe au Chien means Point of Dogs . . ." (Dupre
2001:1) without further elaboration as to why that might be a problem. The
unstated assumption is that it is embarrassing and that this fact should be
obvious to all. That it was a prime motivator for Ledet is clear from statements made by those who knew her ("alle trouvait que c'était une insulte")
and from published material about her in the *Advocate*, which references
"degrading stories about lost dogs" and calls *Pointe aux Chênes* more "genteel"
(Associated Press 1986). A weak supporter of *Chênes* in my own interviews,
though still attributing the sentiment primarily to Ledet, suggested that "Ça
semble it feels better, it sounds better. I guess."

It is not unusual for dogs to be regarded as dirty or undesirable in some
other manner—consider the effect of calling someone a dog, for example,
in English. That said, it is also hardly unusual for place-names to involve
dogs in some way: a search of the GNIS database for American locations
using *dog* brings up nearly seventeen hundred results (compared to more
than two thousand for *oak*) (US Geological Survey 2017a, b)[7] and includes
places named Dog River, Dog Creek, Dog Bay, Dog Island, Dog Head, and,
notably, a Bayou de Chien in Kentucky. Bayou de Chien does not seem to

be suffering the same identity crisis as Pointe aux Chênes. The GNIS records only one variant name for it: Bayou *du* Chien (US Geological Survey 2019a). The *Chien* does not seem to be the element in dispute (nor, presumably, given the absence of recorded variants, was it historically, when residents spoke French).

It is, moreover, also perfectly possible for places to bear much more embarrassing names and to have those names upheld despite the call by some in the community to change them. Such is the case of Dildo, Newfoundland, for example, which has maintained its name despite at least one attempt to change it because enough of the community enjoys the cachet that it brings the town, however covert that cachet may be (Jennings 2013; Levin 2016). Monmonier (2006) documents a host of embarrassing and offensive place-names in the United States, all far worse than *dog* in their denotation, many of which have maintained their names (though his work would seem to suggest that there exists a threshold of offensiveness at which point the community at large gladly supports a name change).

But of course, in this case, the name is *not* embarrassing to *all* the residents of the community. A consideration of historic social forces in Louisiana may explain the apparent split along ethnic lines. Until the mid- to late twentieth century, rural Louisiana francophones suffered from serious prejudice at the hands of anglophones and wealthy francophones. The term *Cajun*, though today worn with pride and bearing a sense of cool, was a fighting word for much of its history. It was applied indiscriminately to all rural white francophones, generally pejoratively, regardless of their degree of Acadian heritage, almost as soon as anglophones arrived in the state (Brasseaux 1992). In my research I came across a few people who remembered avoiding that term in their youth, though they remarked that today it was possibly even prestigious. Walton (2017a) recalls a more pronounced bristling at the term, especially among older residents, in Bourg in the early 1990s. These residents also very strongly favored the *Chênes* variant, often complaining that they had been mocked as "a bunch of dogs" for living on or near the bayou/town with that name. One particularly acerbic resident pointedly asked, "You know what they call female dogs, don't you?" These residents would have felt strong pressure to prove that Cajuns (if they were going to be called that) were intelligent, cultured people, and insisting on the *Chênes* variant (which had long existed) may well have been one way to do so. Social class is a possible motivating factor here, too.

Indians, meanwhile, suffering far greater prejudice simply for their (alleged) racial makeup, had worse problems to worry about than whether someone was considering them lower class. The Indians' identity *as Indians* has been challenged by parish officials and other residents for at least one hundred years. Claims that they are "tri-racial isolates" descended from white, Indian, and black ancestors and therefore not entitled to an identity as Indian have plagued group members for much of their recent history and have resulted in the application of a deeply offensive epithet.[8] A prime example of the discrimination they faced historically, especially from people in positions of authority, comes from Henry L. Bourgeois (1938), whose master's thesis includes an entire chapter, "So-Called Indians," that constitutes an angry screed in which he decries them as "pariahs" (64) who "loaf" nine months of the year (65) and who have resisted all good attempts to access public educa-tion (63) (though he also documents, contradictorily, their "ridiculous" [63] demand to be allowed access to white schools [66] and their "never ceasing visits to the office of the school board" [66]). He goes on to accuse them of occupying an "imaginary racial zone standing midway between the whites and the blacks" (64) and of seeking "social or racial parity with the whites through the acceptance of their children in the white schools" (66), actions he deems to be examples of their "illusions and social aspirations" (66). He later refers to a parent demanding that his children be educated alongside white children as "a pompous blackamoor if ever there was one" (69) and characterizes the community as a whole in additional unsavory terms, not-ing that it has "few, if any, of the earmarks of [its] boasted ancestry" (65). Bourgeois was superintendent of schools at the time.

CONCLUSION

Even before federal recognition became an option in the 1970s, the Indians' struggle to be recognized as Indians overlaps with the Cajuns' struggle to prove themselves worthy of respect from other white people. This set the stage for the events of the late twentieth century. Both groups' identity strug-gles played out in their creation of place via naming, with different results: some people, most often Cajuns, seeking prestige, seized on the more bucolic *Chênes*; others, most often Indians, seeking basic recognition of their ethnic identity in a context of scathing prejudice, seized on the more frequently used and covertly authentic *Chien*. This may also explain the nuances found in my research: individual Cajuns may well have had less exposure to mockery from outsiders than others, or that mockery might not have fazed them. It may even have made them rebel in response. Likewise, individual Indians may have been more susceptible than others to the suggestion that dogs were in some way degrading and that therefore coming from "Dog Point" was to degrade their language and identities. The associations function on an individual level but with group consequences. The stories directly invoked to justify the names are essentially the same; it is clear that people are arguing about the same *place*. If "what people make of their places is closely connected to what they make of themselves" (Basso 1996b:7), then we have instead competing goals of authenticity (Chien) and gentility (Chênes). The stories that the place-name recalls will be either authentic (Chien) or genteel (Chênes). For both groups, the associations their place invokes reflect on their personal identities not only as Cajuns or Indians but as Bayou residents. As such, are they genteel or resistant to the hegemonic culture that has long denied them dignity? The dispute over the name is a product of fact that placèdness (Silverstein 2014)—in this case, an identity linked to the Bayou—both linguistic and space-based, underlies other identities. The next chapter illustrates how Bayou identity, constructed aurally, is strong enough to survive well after residents have shifted to another language.

Chapter 7

LANGUAGE SHIFT AND THE CONTINUED IMPORTANCE OF FRENCH TO BAYOU IDENTITY

Living in the Lafourche Basin alongside the aging francophone population are a large number of people for whom French is no longer a language of daily use or even one that they control with any real capacity. The shift to English has resulted in younger generations that are generally unable to speak French, though it is clearly part of the background of their world. It is evident that English has become the language not only of everyday life but of the expression of identity.[1] Indeed, a few interviewees noted that they hear differences not only in the French of neighboring bayous but in their English as well—though it was often unclear whether this really meant that they could tell via their English that they were native French speakers. This English is influenced by French phonologically, syntactically, and prosodically (Scott 1992; Cox 1992; Walton 1994; Dubois and Horvath 1998a, b, 2000, 2002, 2003a, b, c) and is peppered with French words and expressions, such as *couillon* (stupid) or *envie* (a desire, a wish) (see also Cheramie and Gill 1992). It is also clear, however, that French continues to play an important role in place-making beyond simply influencing the English spoken.

In addition to denying or minimizing interparish variation in French, people frequently reported that the way to identify someone from the lower bayous was that they "talked flat." Walton (1994) suggests that this term may be used as a consequence of the French-based intonation pattern found in English, in which most syllables receive the same weight and only the final syllable of what in French would qualify as a "groupe rythmique" is stressed.[2] I did come across a few instances of people who called their French *plat* (flat), but it is possible this term was carried over from English into French. That said, Walton (1994), citing personal communications, establishes the term's use in both nearby Assumption Parish and on the prairie in the town of Eunice, suggesting that the word enjoys fairly wide use. In any case, I was less concerned with the origins of the term than with its potential to indicate group membership, to index bayou identity. Consequently, in 2012 I followed up on my dissertation work with a map-drawing exercise on perception based on those conducted by Preston (e.g., 1986, 1996, 2011) and others. Though the exercise asked about English, it ultimately showed that French continues to play a large role in the construction of place in Terrebonne-Lafourche despite (or possibly because of) the language's decline. Even beyond French's role in shaping the English of South Louisiana, however, the symbolic importance of the language carries on, well beyond the point at which it is still the language of everyday life for most bayou residents.

MAP-DRAWING SURVEY

The map-drawing survey consisted of a blank map of Louisiana featuring only parish names and boundaries (figure 7.1). My students and I set up in public places around Houma (the parking lot of a Rouse's grocery store frequented by lower bayou residents, for example) and stopped passersby to ask them to participate in a survey on language. We then asked them to show us, using lines or shading and including labels or other information if possible, where people spoke differently in Louisiana. On an attached page, we recorded the respondent's age and gender, the town in which they had spent most of their childhood, their current residence, and their race/ethnicity (presented as "Which of the following would you call yourself [check all that apply]?" followed by a list of options consisting of both racial and ethnic labels including but not limited to White/Caucasian, Black/African American, Asian, Cajun, Creole, Acadian, Houma, Vietnamese, and Italian). When they had completed the survey, we conducted a short interview asking them to discuss their maps and posing any questions we had about them. We also made a point to ask about the term *flat*, whether or not they had included it on their map, asking them to define it and, if they hadn't already included it on their map, mark where people spoke "flat." We made notes on the maps and forms documenting their answers.

We asked about English instead of French, and although such exercises are usually conducted with university students (generally students enrolled in researchers' classes), we instead surveyed members of the public, selected by their presence in public places. This removes a certain bias that is present when university students are surveyed: in such contexts, the results may be influenced by various factors related to the classroom context, such as the possibility of receiving extra credit, (unfounded) fears that grades may be

Figure 7.1. Map used in map-drawing exercise.

affected, the fact of surveying only those who have chosen to pursue years of additional study beyond high school, and so on.

A total of thirty-three people produced usable maps for the survey. All of the respondents were native Louisianans, and they came from the bayou region (Terrebonne, Lafourche and Assumption Parishes) or from the prairie parishes of Lafayette, St. Martin, St. Landry, and others . Three outliers— two from the greater New Orleans area and one from Natchitoches—also completed the survey. Three more declined to give their regions of origin, so their answers had to be discarded for the purpose of comparing the regions, though their responses were included when analyzing the group as a whole. Somewhat surprisingly, given the level of metalinguistic awareness in South Louisiana, we received a number of maps with minimal writing whose authors professed to not know where people might speak differently (which is a functionally different response than that from the person who confidently asserted that everyone spoke the same across the state). We also received a number of refusals to participate on the same grounds (i.e., that they did not know and thus could not answer). On the other end of the scale, however, we received many detailed answers from people who labeled their maps, often with negative nicknames (most often, *redneck*) for the residents of various areas. Figures 7.2–7.5 present a few examples.

A few important findings emerge from the survey:

1. The most frequently indicated division was a north-south divide, present on nineteen maps (58 percent). This division was indicated with a line or via labels or inferred simply because only northern regions were indicated on the map—i.e., people circled places they felt were different from their own, which happened to be in the north. When people actually drew lines (thirteen maps), they varied little in placement and were found approximately at the level of the "ankle" of the Louisiana boot (see for example the *imaginary line* in figure 7.5). The line corresponds roughly to the northern reaches of the French Triangle (one respondent traced nearly perfectly the northern boundary of Acadiana). South Louisianans clearly do NOT consider themselves southern—or at least not southern like any others (the Confederate flag appears too often in South Louisiana—especially rural South Louisiana—to claim no affiliation at all). Possibly identification as southern is situational or intersectional: they may well have labeled themselves southern if presented with a map of the United States instead of the state of Louisiana alone. In any case, the north of the state is often labeled *southern* (somewhat incongruously, the South is in the north), *redneck*, *hillbilly*, or, amusingly, *their own situation* (see figure 7.4). *Drawl* also appears twice and *twang* once. These labels and value judgments like *snotty*, which appears on one map, suggest a level of disdain for North Louisianans and by extension other southerners (though *Texas* or *Texans*—the butt of many disdainful jokes in Louisiana—appears several times as well; Texans are apparently also not typical southerners, though they are also not like South Louisianans). This finding confirms Preston's (1986) conclusion that the most disparaged regions (in his research, the American South and the Northeast) are also the most often identified. In this case, since the task involved a map of Louisiana given only to residents of South Louisiana, it would perhaps be more accurate to say that people are acutely aware of just where the division between "us" and a disparaged "them" is located and they are firmly drawing a boundary around their home territory. Acadiana is also a well-known and well-defined cultural region, so perhaps it should not be too surprising that residents mark it off when asked to do so on a blank map.

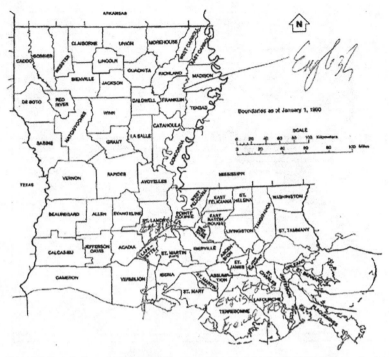

Figure 7.2. Example of a completed map. Northern part of the state is labeled "English"; St. Landry, Terrebonne, and Lafourche Parishes are labeled "English and French."

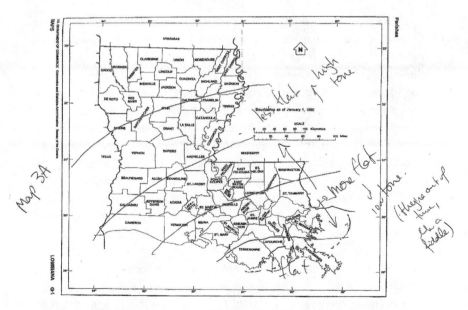

Figure 7.3. Example of completed map with gradations ranging from "Less flat/ high tone" to "more flat/low tone" ("they're out of tune, like a fiddle").

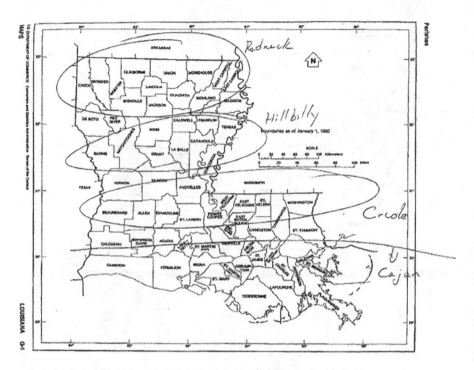

Figure 7.4. Example of completed map labeled Redneck, Hillbilly, Creole, and Cajun from north to south.

2. Five maps have concentric arcs or lines separating regions, indicating a gradation. On these maps, the gradation is in the direction of the Gulf (on two an arrow points toward it), with text indicating that as one approaches the Gulf (i.e., lower Lafourche Parish), speech becomes "thicker" or "more flat" (defined by one as "out of tune, like a fiddle") (figure 7.3). On one map, the progression runs from *Redneck* through *Hillbilly* and *Creole* and ends in *Cajun* (figure 7.4). On that map, *Cajun* appears over the southernmost quarter of the state. It is unclear whether all the labels were meant to be pejorative on some level, but given that neither *Cajun* nor *Creole* are words that cause insult anymore (though *redneck* and *hillbilly* are still quite contentious except perhaps when used by insiders in reference to themselves), it seems fair to interpret the progression as a gradation that runs from most to least disparaged.[3] Accents may be *thick* or *flat,* but speakers are at least not rednecks.

3. The north of the state is also labeled *English* (for example, figures 7.2 and 7.5). On several maps, *English* is given as a counterpart to *Cajun* and *Creole* (or sometimes nothing) in the south. Given that the language traditionally

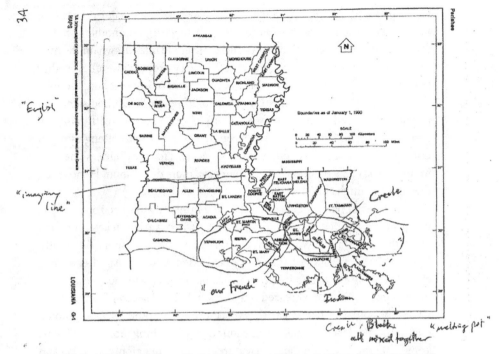

Figure 7.5. Example of completed map. "English" in the north, and below an "imaginary line" are "our French" (over the prairie parishes and extending south to the Gulf); "Creole" (over Orleans, St. Bernard, and the River Parishes); "Indian"; and "Creole, Black all mixed together 'melting pot'" (presumably indicating the coastal parishes of Terrebonne, Lafourche, and Plaquemines).

associated with both Cajuns and Creoles is French (of some variety), I interpreted those labels to refer indirectly to French when I came across them on maps, especially if they were part of a compound such as "Creole influence."

4. The most striking result was the frequency with which people referenced French, including *Cajun* and *Creole*, in their answers. More than half the maps mentioned French in some form. Examples include *French twist, Cajun influence, Bilingual French, French and English,* or simply *Creole/Cajun/French.* Eight people (25 percent of respondents) noted that French was still spoken in some way, whether bilingually or not. One map, for example, is split into a north/south general divide, with the north labeled *English* and the south *our French* (figure 7.5). Another once again labels the north *English* and labels several southern parishes *French & English* (figure 7.2). Some people seemed to think we were actually asking about French rather than English, despite our oral instructions and the written directions on the top of the page that made it clear that we were interested in English: on one map,

the respondent had labeled the southeast *Acadian, not like our French* (the north was labeled *own situation* rather than *English*); on another, I noted, following the discussion, that it was unclear that the lines drawn on the map weren't actually about French (and that they very probably were). A further ten surveys referenced a French/Cajun/Creole *twist* or influence or described the language as *more* or *thicker* Cajun/Creole/French. South Louisiana was frequently labeled—on a survey explicitly about English—with *bilingual French, French and English,* or most strongly, *Old time French.* Roughly equal numbers of people across both regions (54 percent for the prairie, 56 percent for the bayou) only mentioned French or gave more detail about French than they did the English in the north.

A number of participants provided only minimal answers. For example, two people answered that they did not know where variation lay more than to say that they'd met people from the northern areas who spoke differently. Beyond that, they didn't have enough experience to say. Another simply circled the entire state and labeled it *Louisiana English,* with no further comment (figure 7.6). A third placed one tiny circle inside each of Terrebonne, Lafourche, and St. Charles Parishes but included no other label and noted nothing on the form (because discussion revealed nothing more than "I don't know") (figure 7.7). A similar map features a small x in Lafourche Parish and no other mark or explanation whatsoever. If we eliminate such minimal responses, the percentage of people who mention French in each region rises to 80 percent and 75 percent respectively. The number of people who *only* mention French, to the exclusion of other labels except to distinguish it from other languages (i.e., their maps are labeled only with *French* or *Cajun* versus *English* and in one case *German* [around Rayne/Roberts Cove, where until recently German was still spoken]), is higher on the bayou than on the prairie (50 percent versus 30 percent).

In any case, most respondents were mainly concerned with variation in South Louisiana: the number of people who gave minimal to no information about the north of the state (beyond simply labeling it *English* or *Rednecks*) but detail about the south is roughly equivalent across the regions, at 75 percent for the bayou and 70 percent for the prairie. There are several possible explanations for this phenomenon: (1) people legitimately don't know about variation outside of South Louisiana, with bayou residents knowing slightly less than those from the prairie, who are closer to it and may therefore have more experience with it; (2) there really is no significant variation north of the line, or (3) people did not think we wanted to know about variation outside the francophone sphere. Though minimal information on English

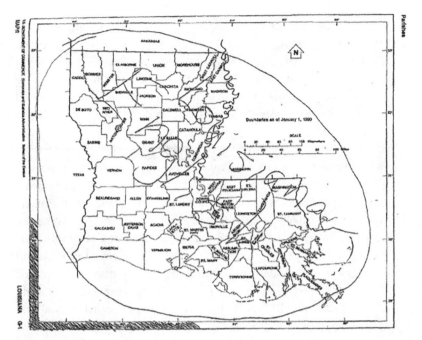

Figure 7.6. Example of a minimal response; the whole state is circled and labeled "Louisiana English."

Figure 7.7. Example of a minimal response: three small circles appear inside Terrebonne, Lafourche, and St. Charles Parishes.

in North Louisiana exists to date, Abney (2019) suggests that significant variation exists in the region. The third explanation is the most plausible, given the clear misconception that the exercise was about French and not English. I encountered the same phenomenon in 2010 when conducting videotaped interviews about Louisiana English: two respondents completed the interview in French despite all efforts to redirect them into English. There was no convincing them that English was of interest.

Most interesting in the phenomenon of respondents focusing on French rather than English is the fact that similar numbers of people from the bayou and elsewhere in the state mention French on their surveys (bayou residents are more likely than others to use the term *Cajun*, but others are more likely to mention *Creole*, likely because the only interviewees who identified as Creole were from outside the bayou), and people associate French with their own homes at roughly the same rate: 80 percent of those from the bayou and 82 percent of prairie residents who mentioned French in some way placed the label on their own home region. Conversely, outsiders are more likely to associate Lafayette with French: 60 percent of bayou participants labeled the prairie *French* (or *Cajun* or *Creole*) (sometimes in addition to their own region, occasionally exclusively), but only 45 percent of outsiders did the same for the bayou region (though that group includes one person who labeled only Lafourche Parish and called it *Strong French*. In short, people generally are convinced that their own home region is more French than anywhere else in the state, though bayou residents are more willing than those from the prairie to concede that other areas might also be French (or might be French instead). This likely has much to do with the fact that the Cajun revival movement, including CODOFIL's offices, is headquartered in Lafayette. While residents are often unaware of CODOFIL, it has nonetheless played an important role in disseminating images of Cajun identity and promoting Cajun culture.

CONCLUSION

These exercises show above all the strength of the connection of French with bayou identity. The survey asked people to discuss English; instead, we often received answers about French. A quarter of respondents not only spoke of French influence or a French "twist" but labeled their maps in a way that suggests a contrast between English (in the north) and Cajun, thereby implying that the latter is not Cajun English but rather French or at least some hybrid of the two (as labels like *French and English* imply). This suggests that French is *enregistered* (Agha 2003; Johnstone, Andrus, and Danielson 2006) in Terrebonne-Lafourche. Johnstone (2013; see also Johnstone, Andrus, and Danielson 2006) explains that enregisterment is a process by which linguistic features that once passed unnoticed are seized upon and become associated with social features. This process generally involves the language of the working class (most famously, via Johnstone's work, in Pittsburgh, Pennsylvania), newly exposed to outsiders, who come to realize that the language they speak differentiates them not only from the upper classes in their own community but also from communities living in other places. Consequently, they interpret the diagnostic features of their language as diagnostic of geographic origin. That the language of the lower classes is connected this way is unsurprising: the upper classes are more mobile and consequently less rooted to the land (and, moreover, tend not to work the land directly), and their language therefore reflects this lack of placèdness. Again, these conclusions are the result of ideological positions and may not reflect linguistic reality; the features may not be limited to the boundaries of the place with which they are thus associated.

Terrebonne-Lafourche lacks stark class distinctions. To be sure, some people are wealthier than others, some have more prestigious jobs, and some have obtained higher levels of education. But in such a community, these class markers, while present, are in many ways less important than they would be in a large city. The size of the community means that your boss is often your neighbor (or a relative), your children attend school together, you often socialize with the same people, and so on. The distance between people of different means is not as great as it might be in a larger community where it is easier to keep acquaintances from different contexts separate from each other. In any case, the francophone population generally grew up in poverty, regardless of the wealth that many have accumulated as adults, so the question of class distinction is moot.

Of further note is the average age of respondents: bayou residents, with an average age of forty, were seventeen years younger than their counterparts on the prairie. None of the respondents claimed to speak French, but it is likely that the most elderly among them may have been at least very weak semispeakers. In any case, people who do not speak French themselves are telling us about variation not in the language they actually speak but rather in the heritage language of their region. The ideology holds that French is spoken; that it is stronger in home regions suggests a connection of the language with a place (for both regions, of course) that outlasts the language's vitality. If anything, the language may have become enregistered *because* of language death. In this case, the encroachment of English replaces population mobility in exposing people to other ways of speaking. French is therefore more authentic, the *real* speech of this region.

In discussing place-based identity in Greenland, Nuttall (2001:64) notes,

> People in Tasiusaq and Aappilattoq have a strong sense of local culture, continuity and tradition—in itself this stands in stark contrast to, possibly masks, or even helps people cope with and adjust to social and economic changes. An obvious example is that people will say quite proudly that they live in a community of hunters and fishers, yet the mode of production that characterizes the hunting way of life and the technology used today is substantially different from that used say fifty years ago. Yet for people in Tasiusaq and Aappilattoq, the hunting "tradition" is alive. This hunting tradition and other traditions are alive in the sense that traditions are common modes of practice. Traditional activities are what people usually do, they are customary, common practices, ways of acting and ways of living one's life. As such they connect the past with the present.

My map-drawing exercise reveals precisely the same phenomenon in Louisiana, though the cultural practice in question here is the use of the French language. Elsewhere (Dajko 2018) I have shown that French remains alive in the imagination for residents of both the countryside and New Orleans, even when people no longer speak the language (over 90 percent of New Orleanians are monolingual English speakers [US Census Bureau 2010b]). Similarly, Lindner (2019) and Valdman (2007) find French functioning as an identity marker despite a lack of fluency in the language. Here, we see that people are more likely to say their own home region is francophone than they are to label any other region so, and the relative youth of respondents in Lafourche suggests that the attachment to French continues even for younger generations with less frequent exposure to French. This is, at least at first glance, somewhat surprising, given the central role that Lafayette—i.e., not their home region—plays in the French language revival movement. It was also predicted by Walton (1994:93), who noted that while English was becoming the language of daily communication, French would remain the symbolic language for the community. Such is the case because place has been mapped onto the language. The next chapter explores the parallels between the way that place is constructed physically and aurally via an examination of the stories told in the construction of place and the discussion of the disappearance of both land and language, confirming that place maps onto the landscape and the soundscape in analogous ways.

Chapter 8

OMENS FROM THE PAST,
WARNINGS FOR THE FUTURE

Storytelling, Place, and Identity

"My daddy was the first man who knew! About the storm. That devastate the people of Leeville." So began the story of the ordeal of the people of Leeville in the New Orleans hurricane of 1915, as told to me on one (slightly less) stormy summer night.[1] I hadn't planned to interview Ignace Collins; he did not live in one of the communities I was targeting. But he had requested the interview via a friend who told him about my project. And so, my fieldwork assistant Roland and I showed up to talk with the nearly ninety-year-old Galliano man. We arrived at his house at 7:30 p.m., as a light rain was falling; Ignace was eagerly expecting us.

In the end, Ignace did not do my translation exercise and in fact conducted the vast bulk of the interview in English. He was consequently not included in my study of the language. However, he nonetheless provided an important oral history that informed my understanding of the construction of place in Terrebonne-Lafourche. Ignace's account of the destruction of Leeville, which I have reproduced in nearly its entirety here, was emblematic of the kinds of stories that bayou residents tell in creating the Bayou region as a place. Basso (1996b:5) writes that place-making via storytelling involves both remembering and imagining: "What is remembered about a particular place—including, prominently, verbal and visual accounts of what has transpired there—guides and constrains how it will be imagined by delimiting a field of workable possibilities." What is remembered on the bayou is a series of tales of repeated destruction via natural disaster—of instability—that goes back generations. The accounts often also remind listeners of the importance and previous dominance of the French language: though Ignace told his story in English, it was peppered with French words and expressions, many of which he translated, and Ignace took pains to point to the importance of French as he spoke. Stories like these emphasize the precarious nature of life on the bayou, highlighting the fact that both land and language are at risk of disappearing with every hurricane that strikes.

Ignace, it turned out, was a force of nature himself. He burst through the door to greet us with great gusto, eyes glinting with enthusiasm. He had two tape recorders sitting out on his kitchen table. The room itself was piled high with cigar boxes containing photos, cassettes, and all manner of knickknacks, among them two of the miniature churches he was locally famous for making (there were fifty more around the house and another hundred in the garage, he informed us). Prior to our arrival, he had been listening to tapes of himself that he had apparently recorded about twenty years earlier to document his family's history. When I asked what was on them, he hit the "play" button on the recorder nearest his hand, producing a burst of accordion music. When the song ended, his voice happily chirped out of the machine, sounding much younger and stronger, giving us the name of the song. The tape immediately switched to another song. We were cheerfully informed at the end of that one that we had just heard "J'ai Passé devant Ta Porte." Today's Ignace talked over the music, in French, telling us he'd learned to play from his father, who had learned from his father, though both sons reported lesser abilities than their own fathers. And then he decisively announced his intention to get the interview started and to tell us his most important story.

THE DESTRUCTION OF LEEVILLE

It was about 4:30 in the evening. My daddy was just leaving his house to meet his friend at the little grocery store at the East Canal, to meet to shoot the bull. You know what I mean? Talk! They used to do that when they was not at seining[2] or something like that and they had something to talk about. They would meet there, a bunch of young, young student—my daddy was twenty-two—and, uh, and he came right there, he says, and he look up the bayou. My daddy saw a little white boat coming, all painted white, a little boat. Coming in. And he say I'm gonna go and walk—they had to walk a good way before they get to the road. So then the boat came to him, he saw my daddy on the wharf, he stood on the edge of the little boat and he land on the wharf.

He says to my daddy—and he says that in French!—he says to my daddy, he say, "*Jeune homme*. Young man. *Jeune homme, mon, je viens de Lockport*. I come from Lockport!" But in French, he says that. He says, "I came to warn the people of Leeville to get their family on their boat and get as high as they can up the bayou, because there's a big storm coming to hit y'all over here." He says, "It's 4:30, and," he says, "it's time for me to go back home. I got a long way to go to reach Lockport." Turned around he left.

Established by refugees (including Ignace's family) after an 1893 hurricane destroyed the town of Chenière Caminada, adjacent to Grand Isle, Leeville today is a handful of houses located about ten miles below Golden Meadow. Ignace was born in Golden Meadow, following the family's relocation in the wake of yet another hurricane in 1915. He explained that when he was fourteen years old, he began to accompany his father and the other fishermen of the community to work. In this way he heard not only the men's personal experiences but also stories passed down to them from their parents and grandparents.

"Do you know how many people, in 1915, the storm of Leeville, out of twenty-five hundred, how many drowned?" he asked us. We did not.

"Seventeen hundred drowned!"

Ignace told his stories in a heavily French-influenced English; it was clear that French had always been his dominant language despite the fact that he had segued into English. Its influence reverberated through his speech, in his pronunciation, his syntax, and his morphology. Despite a recent stroke that had resulted in frequent pauses to recall information, his recounting of the hurricane that essentially reduced Leeville to the handful of houses it is today was dramatic and detailed, beginning with the ominous appearance of the man from Lockport. Of course, the residents of Leeville were less than eager to believe the warning Ignace's father had been charged with bearing.

There was an old man in his house, not too far, he saw the little boat stop and talk with my daddy. He says to my daddy, "Hey, who was that man," he said, "in that boat?" He says, "I never saw that boat, no."

My daddy says, "I'm gonna tell you what he told me. He come from Lockport, and he want us—he want me to pass the people and warn the people of Leeville, because it's a storm coming our way, to get their family in their boat, and take off. What should we do?"

There was no engine in those boat; it was all sail those days. My daddy says the sun was still shining, a nice breeze about maybe ten, twelve, thirteen miles an hour.

The man, he said, "Eh, that man must be nuts. He must be nuts. A storm?! Bah! Listen to that, a storm. He must be nuts. *Il est fou, lui!*"

> Okay, well then my daddy keep on going, and when he reach his destination at the little store, he told the people in there—well, there was about a dozen that was his friend, was there. He start talking and they—ah, they all laugh. They all laugh. "Choa![3] Lockport! He came from Lockport to tell us that? *Oh, non, il est fou, lui*. He's crazy!"

In other interviews, we learned that other signs of an approaching storm include the arrival of frigate birds, who generally live over the water but fly inland when a storm is approaching. In this case, however, the birds were likely still at sea and may have flown in later under cover of darkness. Or perhaps simply no one noticed them. In any case, the man from Lockport was ignored.

> The sun was still shining, the wind [was mild]. But during the night was different. Start blowing. My daddy says the water start coming up, coming up. The next morning at daybreak they had water—in Leeville they had water almost to their knees. *Leurs genoux*. That's how far, in the night, was coming up, coming up, coming up. So, what should you do? They had a boat, but what you gonna do about sail? Wind was blowing about fifty miles an hour, you know. That storm came up all of a sudden.

Roads were virtually nonexistent at the time, and automobiles were not yet in wide use, especially among the poorer people, meaning that having ignored the man's warning, the people of Leeville were now trapped. Their only option was to take refuge in the strongest structure available—in this case, a sturdy camp (a secondary home generally used for fishing or hunting trips; in the lower bayous, camps are often interspersed with residential dwellings and are often indistinguishable from them, at least to the untrained eye).

> The people used to pass with their children: hold their hands, water up to their knees. It's morning now. This is about fifteen hour later than when the man came from Lockport and tell them about the storm coming. So my daddy—his daddy had died, you know what I mean, and he was buried at Chenière. With the yellow fever. So my grandma, Eveline Perrin, she was a widow. My father was, at the time, was twenty, like I said twenty-two, twenty-three years old, and one of his brother—because there was, I think there was nine, nine people in the family—took all the responsibility of the family! His mama was

still living but both of them was seining, or tramailing,[4] or fishing, see, oystering.

So, people was passing through. Says to his mama, "Mama, people still bringing their children, hold their hand and going to the *loge*."

The word *loge*, it mean safety, where you can protect yourself. It was the camp! That the Woodmen of the World had built. It was built eight feet from the ground. The man who bought, who bought the lodge, he bought the land in the camp for the lumber. They would come once in a while and spend a week and then going back to Baton Rouge and wherever. Sportmen, you know what I mean? Just like today people going to Chenière and Leeville.

And my grandma, she says to my daddy, she says, "No, not yet." But she says, "'Tit[5] Jacques"—his brother, who had took over the family responsibility with my daddy, they was the two oldest.

She says that "'Tit Jacques, he's not there. He's still with Uncle Joe." Joe Perrin. Joe Perrin was one of my grandma brother and he was sein-ing; he had a company of seine, seining, for shrimp.

And she says, uh, "If he come during the night . . ."

Now, things was tough now. People was, people were worried, you know, the storm was getting close, close and tough! Wind and wind! One way, one way, squall, squall. Storm! It took a little time to destroy seventeen hundred people out of twenty-five hundred.

And then he said, "Well," he says, "you know what happened, Mama, at Chenière, *hein*? The people, they waited too long."

So he said, "I'm taking the kids with me. I'm gonna bring them to the lodge." Because he knew that was the strongest building in that time. It was built to hold and stand a storm. They knew exactly what happened. So they built that high for the wind, high from the ground, eight feet. Twelve by twelve. There were no piling to build strong cabin those days, just some square pieces, twelve feet. Boulonné [bolted] up there and boulonné down there. Good foundation, because those people had money. People from Baton Rouge, New Orleans, all those place—Alexandria. Everybody would meet, every once in a while they would come and visit their camp.

He said, "I'm taking the kid before the building will be full of people." And my daddy said he took up, he took them by the hand. Madame Alida Degrandile was one of his sister. She had two kids un-der her arm. But before getting inside he passed the kids to the other peoples inside.

The house was full. The house was only standing, standing up people. Some kids was lying down with their mama and father on the floor, sleeping, you know what I mean. I remember the camp, too, but I never went inside like that. Before it was torn down, the camp was, I'd say, I'd say roughly about forty by fifty in size, you know? A good size. It was a camp, it was built for recreation, you know?

He say after he pass the last kid inside to the people inside he turned back his head, turn toward where they came, where he had left his mother at the camp, she didn't want to go now. She was waiting for 'Tit Jacques. He says he was relieved. His mama was coming, was holding her dress over her knee, and she was coming inside. She got in, and they spent the night in the camp for the storm.

It was clear that Ignace had been an accomplished storyteller before the stroke. Even now, he had us captivated. Meanwhile, as we'd been talking, the rain outside had become a thunderstorm. Already we could hear thunder rumbling as he described his father and his family fleeing to the lodge. By now, darkness had fallen, and with the storm outside, the setting was providing a vivid reenactment of the events of 1915.

'Tit Jacques, it transpired, had managed to ride out the storm on a boat with his fellow seiners, lashed down with all the rope they had. They had taken the boat into a place where the oak trees were particularly thick (a contrast to the general lack of trees in the area today).

> Now, at Leeville, my daddy said, the wave, with the wind before it shift, would come and hit the camp from under the flooring. He says, as if the storm wanted to pull them off!

As Ignace described the waves hitting the building from below, a clap of thunder illustrated quite dramatically what it must have been like for the people inside the lodge, and we could hear rain pounding on the roof above us.

> And that those big boulonnés [pilings on which the house rested], those big up and down, them. These twelve by twelve, was supposed to be ten to twelve feet deep inside, so I guess that was to bring it up. And he said, they stand there, so all the people, standing and hold it from breaking. The floor, the weight, they were all standing up, their weight and everything hold it.

They didn't come out, the wind was blowing, blowing the east. East wind, east wind, east wind, and finally, like all storm, it shift. People don't understand that, see? You got the wind on this side, then, after shift, the storm pass over you, you got the wind the opposite way. That's when everything, there was no building left. Those that slept in the lodge—but the lodge stayed there—they saved themselves. Nobody died inside the lodge.

Now, after the storm, that was different. The men used to pass all over the prairie in the marsh everywhere pick up, drag in, drag in, drag in, the drown people, you know what I mean. Kids, mama, daddy, father. And then they would, four, five, six, eight, ten body, and then dig a big hole, and then bury them right in there. Now, like I said they didn't drown, not all of them, not the seventeen hundred. Do you know how many of them got caught in some debris of camp, some lumber, roof, a boat? Wind shift west, everything in the Gulf. Everything went in the Gulf. Nothing to eat. Nothing to drink. Know what I mean? Nobody knew exactly what happened to them.

THE AFTERMATH OF POWERFUL HURRICANES

Ignace's account is strikingly accurate. Most of it can be verified with other records. The New Orleans hurricane of 1915 made landfall near Grand Isle (possibly directly over Leeville) on September 29, with winds building all day and the storm itself arriving in the afternoon and continuing into the night. This matches Ignace's account perfectly, and there is no reason to doubt that his father—whose age genealogical research confirms was in fact twenty-two at the time—and his family took refuge at a solid camp built by the Woodmen of the World, especially since Ignace remembered the building. While Leeville was largely abandoned after the storm, the number of fatalities was nowhere near seventeen hundred, however. That number is closer to the total population of the town of Chenière Caminada prior to the 1893 storm that destroyed it and from which, probably not coincidentally, Ignace's family had come. His descriptions of people surviving (or not) on improvised rafts of lumber, roofs, and so on match contemporary descriptions of that storm (see, e.g., *New Orleans Daily Picayune* 1893a, b, c).

Possibly as a result of the effects of the stroke, Ignace was likely conflating details from stories told to him about his great-grandfather, who immigrated from France as a teenager, married a local woman, and settled with his family at Chenière Caminada. Ignace's father, Philozate, was the son of the family's youngest son and was born in Chenière about eight months before the 1893 hurricane. That storm blew ashore on the night of October 1–2 and claimed the lives of just over half of Chenière's fifteen hundred people. Both of Ignace's great-grandparents were among the dead, along with, Ignace added, two of their daughters (genealogical research shows that a daughter-in-law and two young grandsons also drowned). Their bodies were never found.

The town of Chenière Caminada still exists, but it is only a small collection of houses today. Most of the population moved away after 1893, by and large up the bayou (though some, including some of Ignace's family, seem to have stayed in Chenière or moved a few miles east to Grand Isle). Ignace's father's family appears to have joined those who established Leeville a few years prior to the 1915 hurricane. Philozate told his son about the 1915 storm but was too young to remember the storm that destroyed Chenière and killed his grandparents, aunts, and cousins. Ignace's grandmother (i.e., the mama who didn't want to leave without her son 'Tit Jacques), however, would have remembered it well, and Ignace knew her and reported that she told him stories as well. Ignace may well have also heard about it from other relatives and people old enough to remember—a likely scenario for someone growing up in that area, particularly since Ignace had listened to the stories told by the fishermen. The trauma of losing fully half a community, including a number of family members, was thereby passed down to Ignace. Ignace's grandmother also told him about the 1856 hurricane at Isle Dernière (Last Island)—now Isles Dernières (Last Islands)—which led to the complete abandonment of that settlement; in this case, she would presumably have been passing along secondhand reports from older relatives or community members.

STORYTELLING, FRENCH, AND PLACE-MAKING

Part of Ignace's intent in telling the story of Leeville was to note the clear progression of destruction over time and the repetition of history (Isle Dernière > Chenière Caminada > Leeville). His account strikingly illustrates how the disappearance of the land and the destruction of communities has been a fact of life in the lower bayous for generations. Stories of destruction are carried forward, and recent experience compounds these stories, making the cycle of destruction and displacement not only remembered and imagined but relived by each new generation, though given twenty-first-century technology, modern events may result in the loss of fewer lives. If what is remembered about a place guides and constrains future imaginings of it, then what is available for residents of the coastal marshes is a frame of destruction. The fact of land subsidence provides the backdrop to life in coastal Louisiana; the stories drawn on for place-making in lower Lafourche include stories of disappearance, of places that no longer exist (or that exist only as shells of their former selves), of places whose boundaries have shrunk. My interviewees repeatedly stressed the constant abandonment of older areas of settlement. Other residents had similar stories to Ignace's. Accounts of two historic storms from Wenceslaus Billiot, a resident of the Île à Jean Charles whose father was born in Pointe aux Chênes, demonstrate the same components as Ignace's account:

(8.1)

WB: L'ouragan de dix-neuf-cent vingt-six, mon pape a dit que XX passé sept pieds d'eau, pas ici, euh, boutte d'en bas. XX boutte là-bas x icitte-là XX dans les cinq pieds d'eau ici parce que c'est haut ici. Mais boutte là-bas c'était, c'était plus bas.	WB: *The hurricane 1926, my dad said that there was more than seven feet of water, uh, not here, but on the ridge below. Here there were about five feet of water because the land is high, but the ridge down there was much lower.*
R1: Ouais, en vingt-six c'est—la terre était plus haut ici.	R1: *Yeah, in '26 it—the land was higher here?*
WB: Oh ouais. Cet ouragan là ça reclame et puis il a tapé Houma. Il a été à Shreveport.	WB: *Oh, yeah. That hurricane came in and then it hit Houma. It went to Shreveport.*

Wenceslaus's family took refuge on boats (which float on top of the storm surge) rather than in a strong building, but this strategy's success relied on the presence of trees and land to absorb the wind and storm surge:

(8.2)

WB: Mais, dans ce temps-là, c'était pas comme asteur. Uh, un ouragan menait et y avait plein des bois.	WB: *But in those days, it wasn't like it is today. Uh, a hurricane came and there were lots of trees.*
R1: Ouais.	R1: *Yeah.*
WB: Plein, plein des bois ici (sus l'île). Et là les drigailles menait,	WB: *Lots, lots of trees here (on the island). And the debris came,*

ça arrêt—ça arrêtait l'eau, ça
arrêtait le courant.
R1: Ouais, okay.
WB: Avec, uh, comme ça eusse
avait pas, eusse avait pas du
courant comme asteur.
R1: Hmm.
WB: Mais asteur y a plus d'arrière.
Y a plus d'arrière. Ça s'en va.

and they stop—they stopped the
water, they stopped the current.
R1: Yeah, okay.
WB: With, uh, so they didn't have,
they didn't have the storm surge
like today.
R1: Hmm.
WB: But now there's nothing back
there. There's nothing back
there. It's all disappearing.

The storm of 1926 was notable, but its death toll near zero. Other hurricanes were more destructive:

(8.3)

WB: Crois pas qu'il y a trop de
monde qu'a perdu leur vie en
vingt-six. Mais le, mais le—je
crois en dix-neuf cent neuf,
là il a plein du monde qui s'a
noyé, en bas sus la Pointe.
R1: En bas du Cutoff, ou ... ?
WB: Ouais, en bas du Cutoff.
Éna de mes parents qui s'a
noyé étou parce que c'est là-
bas où mon pape venait. En
bas de la Pointe, uh, ina plein
qui s'a noyé.

WB: I don't think a lot of people
lost their lives in '26. But the,
but the—I think in 1909, a
lot of people died in that one,
down on the Point.

R1: Below the Cutoff, or ... ?
WB: Yeah, below the Cutoff.
Some of my family drowned,
too, because that's where my
dad came from. At the bottom
of the Point, uh, there were a
lot of people who drowned.

On another occasion, Wenceslaus told the story of the 1909 hurricane in a little more detail:

(8.4)

Ça mon pape m'a conté, ça. Lui
il avait, crois qui'l avait cinq ans,
lui. Il dit en bas de la Pointe là-bas
l'ouragan a menu. Il avait, il avait
du monde qu'a embarqué dans
les bateaux. Inavait un vieux, lui
il est pas embarqué là-dedans.

My dad told me this [story]. He
was, I think he was five years old,
him. He said the hurricane came
to the bottom of the Point. There
were, there were people who got
onto boats. There was an old man
who didn't get on the boats. And

Et il a monté dans un bois, dans une chêne. Et là il a menu que ça a duré assez longtemps qu'il reste il s'a fatigué. Et là il a tombé il s'a noyé. Autrement que ça t'es dans un bateau-là, t'es *alright*. Mais avant ça nous-autres, nous-autres on allait pas en haut. On restait dans le bateau XX / X avait un bateau tu connais / on allais pas en haut. C'est comme ça que c'était, la vie.	*he climbed a tree, an oak. And it happened that it lasted a long time and he got tired. And he fell and he drowned. Otherwise, if you're in a boat, you're alright. But back then we didn't evacuate to higher ground. We stayed on boats. If you had a boat, you know, you didn't evacuate. That's how life was.*

Wenceslaus's story always ends on the same note that Ignace's does:

(8.5)

So, asteur c'est plus pareil. Ina plus de bois non plus. Un bateau, uh, la lame va s'en venir et ça va. C'est plus comme c'était avant. Avant ça il avait des, des lames XX qui menait contre les bois étou, ça arrêtait le courant. On avait plus de courant. Mais asteur y a plus rien du tout. {laughs} Tu pries. Il faut que tu fais route en haut.	*So today it's not the same. There are no trees anymore, either. In a boat, a wave comes and goes, and it's okay. But it's not like that anymore. Before there were, there were waves that came into the trees and all, and that stopped the surge. We didn't have a surge after that. But today there's nothing at all. {laughs}. You pray. You have to hit the road to the high ground.*

While such stories aren't always as dramatic as Ignace's or Wenceslaus's, they are the backdrop to most people's personal histories. Storms are often the catalyst for movement across the region. In the following example of a very limited hurricane narrative, no one dies and no dramatic details are given, but the effect is that the family is displaced—in this case to the east and then to the north, always in search of higher ground. The interview was conducted in Larose, but the interviewee had only been there a short while, having moved into a house owned by her father's aunt; the family had moved first from Pointe aux Chênes (which, she notes, was further down then than it is now) to Golden Meadow and from there up the bayou:

(8.6)

L1: Parce-que la vieille qui restait là dans la maison ici, c'était la tante à mon *pop*.	L1: *Because the old woman who lived here, she was my dad's aunt.*
R1: Mmm-hmm?	R1: *Mmm-hmm?*
L1: Et elle parlait français.	L1: *And she spoke French.*
R1: Et elle devenait d'ici, ou d'en bas aussi?	R1: *And was she from here, or was she also from further down?*
L1: Uh, elle menait d'en bas.	L1: *Uh, she came from further down.*
R1: Ah, okay.	R1: *Ah, okay.*
L1: Ça menait, mon *pop*, la mame à mon *pop* là menait de la Pointe—la Pointe au Chien. *But* c'était pas éyoù ce que le village est asteur.	L1: *They came, my dad, my dad's mom came from the Point—from Pointe au Chien. But it wasn't where the town is today.*
R1: Mmm-hmm. Ah, c'était plus en bas du Cutoff et—	R1: *Mmm-hmm. Ah, it was below the Cutoff and—*
L1: C'était en bas du, dessus le Cutoff là-bas.	L1: *It was below the, below the Cutoff there.*
R1: Ouais okay.	R1: *Yeah, okay.*
L1: Et *so* c'était, um, eusse avait une sucrerie, ça, ça plantait des cannes là étou. Et là c'est là éyoù ce qu'eusse, eux-autres devenait.	L1: *And so it was, um, there was a sugar mill, they planted sugarcane there, too. And that's where they came from, them.*
R1: Huh.	R1: *Huh.*
L1: *But* ça c'est XX.	L1: *But that's XX.*
R1: Équand ce qu'eusse a grouillé ici?	R1: *When did they move here?*
L1: Hoh! *Lord.* Je m'en rappelle pas. C'est après, apl—après l'ouragan de dix-neuf cent trente-sept, je crois.	L1: *Hoh! Lord. I don't remember. It was after—after the hurricane of 1937, I think.*

While either the date of the hurricane is inaccurate (there were no hurricanes in South Louisiana in 1937) or it was in fact a tropical storm, what is important is that the family had to move because the land had become uninhabitable, at least during storms.

For some older interviewees, the stories were not even about family history; the interviewees themselves were the protagonists. One monolingual French speaker explained that he had grown up in Pointe aux Chênes but below the Cutoff Canal. While many people only spent winters there, his and several other families lived in a permanent settlement miles below the lowest boundaries of the current town. The town was unconnected to the rest of the parish by road and could be reached only by boat—which back then meant a pirogue. The settlement had been abandoned too many years ago to count, however; while a small ridge of land was still there, he told us it had long ago given over to saltwater and long grass, and he wasn't sure when he had last lived there. That it was still home to him was clear, however: he volunteered the information, as did all the others who lived there, before I asked for details and in this case even before I asked where he was born.

Many of the stories were also very recent. One interviewee told the story of a niece who had lost everything in a storm and moved to Houma, as had her daughter. Another niece had moved to Mississippi. An interviewee on the Island lamented the fact that his father hadn't settled further up the bayou as the ground beneath his house was now washing away, and he would soon have to move. The constant abandonment of older areas of settlement and the movement of towns upstream was a recurring theme. For example,

(8.7)

Dulac habitude d'être loin en bas.   *Dulac used to be way down the bayou.*

(8.8)

R1: Et t'as été énée et élevé entour ici?	R1: *And were you born and raised around here?*
L1: Ouais. J'ai tout le temps resté icitte.	L1: *Yeah. I've always lived here.*
R1: Ah, ouais. Toute ta vie. T'as jamais resté dans une autre place?	R1: *Really. All your life. And you've never lived anywhere else?*
L1: *No.* Toute ma vie ici. Mais avant ça on restait plus en bas!	L1: *No. All my life here. But before we lived further down [the bayou]*
R1: Ah, ouais?	R1: *Oh, really?*
L1: Plus là-bas en bas.	L1: *Further down, further down.*

(8.9)

R1: Et toi eyoù t'es éné et élevé?

R1: *And where were you born and raised?*

L1: Uh, Pointe au Chien.

L1: *Uh, Pointe au Chien.*

R1: Okay.

R1: *Okay.*

L1: Douze miles en, en bas du chemin de la Pointe.

L1: *Twelve miles below the road at the end of the Point.*

R1: Douze miles en bas de la Pointe asteur?

R1: *Twelve miles below the bottom of the Point today?*

L1: Ouais. Vois XX où le bout du chemin arrive là?

L1: *Yeah. You see XX at the end of the road there?*

R1: Uh-huh.

R1: *Uh-huh.*

L1: Douze miles en bas de ça.

L1: *Twelve miles below that.*

R1: Ah oui.

R1: *Oh, really.*

L1: 'N allait à pied au magasin là-bas.

L1: *We walked to the store there.*

R1: Uh-huh.

R1: *Uh-huh.*

L1: Si la marée était trop basse on allait, on allait en pirogue— uh, pied je veux dire.

L1: *If the tide was too low we went, we went by pirogue—uh, we walked, I mean.*

R1: Huh.

R1: *Huh.*

L1: Si la mer était haute on allait en pirogue. À la pagaille.

L1: *If the tide was high we went by pirogue. We paddled.*

R1: Huh. C'est long!

R1: *Huh. That's far!*

L1: Ouais.

L1: *Yeah.*

R1: Douze miles.

R1: *Twelve miles.*

L1: Ouais. Douze miles.

L1: *Yeah, twelve miles.*

R1: Ça s'appelait toujours la Pointe au Chien là-bas, ou uh ...?

R1: *Was it still called Pointe au Chien down there, or, uh ...*

L1: Ouais toujours la Pointe au Chien de—derrien boutte en bas là.

L1: *Yeah, it was still Pointe au Chien, below the end of the ridge down there.*

In all these cases, it is notable is that despite the movement of more than ten miles, the town still bears the same name. Equally noteworthy in these stories is the inclusion of commentary regarding the language used. Ignace code-switches into French a number of times and stresses at points that conversations were conducted in that language. Meanwhile, the speaker in (8.6) includes the detail that her aunt spoke French. Nearly all of my interviews were conducted in French; to note that someone spoke French—after having made clear that French was often the *only* language spoken by that generation, thus making the note redundant—highlights the importance that the language has for its speakers and simultaneously stresses its disappearance. In fact, just prior to beginning her story, the speaker in (8.6) commented on how refreshing it was to be able to speak French with someone.

STORIES OF LAND LOSS, STORIES OF LANGUAGE LOSS

A perpetual theme running through the stories, then, is change. Rapid, unwelcome change, ending in disappearance. And this disappearance and change are affecting the language just as they are the land. Though language loss lacks the elaborate narratives that hurricanes can solicit (though residents often blamed schooling or even themselves for the shift to English), stories about destruction directly invoke French. The disappearance of the two is also discussed in the same terms: stories told about language parallel those told about land. One frequent approach is to compare the past with the present. Just as we hear about disappearing trees now replaced by open water, we hear that everyone used to speak French, but now there's English everywhere. Just as we hear about muskrats and frigate birds that no longer frequent the area (the former allegedly drowned by hurricanes, the latter disappearing for no known reason), we hear about words that used to be common that no one uses anymore. Whether the change is linguistic of physical, there is a repetition of *dans le temps* (back in the day) and *dans le vieux temps* (in the old days).

(8.10)

L1: Dans un temps, notre français est est est plein des paroles, et c'était pas comme asteur (dans ce temps).

R1: Ah, ouais?

L1: *No, they changed*, tu vois?

R1: Ah, ouais? Comment ça?

L1: En français on use pas les mêmes paroles comme le monde parlait français dans le vieux temps.

L1: *Back in the day, our French was, was, was full of words, and it wasn't like it is today (back then).*

R1: *Oh, yeah?*

L1: *No, they changed, you see?*

R1: *Oh, yeah? How so?*

L1: *In French we don't use the same words like people who spoke French back in the old days.*

(8.11)

L1: Là, icitte-là ça, ça faisait
récolte, dans le vieux temps.
Mon grand-grand-père,
lui il a acheté ça ici—terre
là, icitte-là dans le 1876. Ça
ça donne, cent et quelques
années?

R1: Cent trente.

L1: Plus que cent ans passés. Et
dans ce temps là y avait y
avait pas de, les, tu voyais pas
l'eau comme asteur.

R1: Hmm.

L2: Tu voyais les bois.

L1: Tu voyais des bois. Y avait
plein des bois.

L2: Y en en a plus.

L1: *Oh no*, y a plus de bois asteur.
L'eau salé après tout nous
manger. Et on peut plus
planter.

L1: *Now, here, they used to raise
crops, in the old days. My
grandfather, he bought this
land here, in the, in 1876.
That's, what, a hundred and
some years ago?*

R1: *130.*

L1: *More than a hundred years
ago. And in those days there
wasn't, there wasn't any, the,
you didn't see the water like
you do now.*

R1: *Hmm.*

L2: *You saw the trees.*

L1: *You saw trees. There were lots
of trees.*

L2: *There aren't any anymore.*

L1: *Oh, no, there are no trees
anymore. The salt water ate
them all. And we can't plant
them anymore.*

(8.12)

Dans le vieux temps ça parl—y
a plein du monde qui parlait
français là itou. Mais c'est, c'est
tout après changer.

*In the old days they spo—there
were lots of people who spoke
French here, too. But it's, it's all
changing.*

(8.13)

Dans un temps c'était haut mais
asteur c'est tout calée. C'est bas.
Bien bas.

*Back in the day it [the land] was
high, but now it's all sunk. It's
low. Really low.*

We are further told that neither land nor language can ever be the same
again. In Golden Meadow, an interviewee insisted that the youth would
never learn to speak French as it was meant to be spoken in the area, and in
Pointe aux Chênes an interviewee detailed changes to the Île à Jean Charles
that were so drastic she felt that it no longer seemed like the same place:

(8.14)

No. Le jeune (monde) va pas regarder comment ce qu'on parle. Eusse va parler en français mais tu vas pas les comprendre. Nous-autres, on va se comprendre entre nous-autres, on se comprend, mais eusse, eusse va pas parle—eusse eusse va le—eusse va le parler, mais eusse vas se parler une différent manière que nous-autres on se parle. Tu comprends qui j'après te dire, hein? Eusse, eusse va parler un vrai français. Nous-autres on parle le français cadien, mêlé avec le français en anglais.

No. The young people aren't going to listen to how we talk. They're going to speak French, but you won't be able to understand them. Us, we'll understand each other, but they, they're not going to speak—they're going to—they're going to speak it, but they'll speak it a different way than we do. You know what I'm saying, huh? They're going to speak the real French. Us, we speak Cajun French, mixed with French and English.

(8.15)

Quand nous-autres on était jeune on allait pêcher là-bas à la Poi—uh à l'Île à Jean Charles, on allait pêcher des crabes étou, et y avait plein de la terre là encore dans ce temps là. Mais asteur, uh, c'est tout de l'eau tout partout. On a été *ride*—ma fille m'a amené *ride*, um, l'année passée. Et ça semble plus à la même place seulement. C'est, uh, c'est plus pareil du tout. Tu vois la Pointe étou les bois c'est tout crevé? C'est, uh, c'est *put it that way*: c'est vilain. {*laughs*} C'est pas le, c'est, um, plus du tout pareil comme c'était.

When we were young, we used to go fishing down there below the Poi—uh, on the Île à Jean Charles, we went crab fishing, too, and there was still lots of land there back in those days. But now, it's all water everywhere. We went down there—my daughter took me down there last year. And it doesn't even look like the same place. It's, uh, it's not the same at all. You know on the Point also, all the trees are dead? It's, uh, it's put it that way: it's ugly. {laughs} It's not the, it's, um, not at all like it used to be.

Several interviewees gave timelines, whether in terms of years or generations, for the disappearance of either the land or the language:

(8.16)

R1: Eusse dit que ça va durer encore combien longtemps, la la terre en en bas ici, avant que ça va être mange aussi par la—

L1: Oh! un autre vingt ans je [pourrais] plus dire arien du tout qui va rester ici.

R1: *How long do they say it's going to last, the, the land down here, before it also get eaten by the—*

L1: *Oh! Another twenty years, I'd say, before there will be nothing left here at all.*

(8.17)

R1: Et le monde parle pas français dans les restaurants, ou …

L1: Non. C'est malheureux, *though*, parce que le français après s'en aller. Quand XX à nous-autres va s'en aller y en aura plus du français.

R1: *And people don't speak French in restaurants, or …*

L1: *No. It's sad, though, because French is disappearing. When we're gone there won't be any French anymore.*

(8.18)

On après perdre notre langue français ici à Jean Charles. Um, dans leur familles, les neuveux et les nieces, les plus vieux, appris à parler français, mais, uh, leurs enfants à eusse, il y en a qu'a appris un 'tit brin le français mais il ont pas montré, uh le français de trop. Ça fait que, um, le, la langue français est comme une generation, deux générations plus bas que moi. Après ça, c'est, c'est tout en anglais.

We're losing our French here on Jean Charles. Um, in their families, my nieces and nephews, the oldest ones, learned to speak French, but, uh, their kids, some of them learned a little bit but they didn't teach them, uh, too much French. So, um, French is like a generation, two generations below me. After that, it's, it's all in English.

(8.19)

L1: Ça va jamais continuer à parler le français. Ça ça va être
une affaire qui va—tu pour-
ras dire *goodbye.* XX dans
les à menir [avenir], parce-
que comme tout (le) vieux
monde va p—va se mourir,
c'est juste du monde comme,
uh, toi là qu'après essayer
d'apprendre en français.

R1: Ouais.

L1: Et t'aurais personne d'autre à
parler avec, qu'autrement tu
trouves un vieux, ou quelque
chose comme ça.

L1: *They'll never keep speaking*
French. It's going to be a thing
that's going to—you can say
good-bye to it. XX in the
future, because all the old
people are going to die, and
it's only people like, uh, like
you who are trying to learn to
speak French.

R1: *Yeah.*

L1: *And you'll have nobody else*
to talk it to, unless you find an
old person or something like
that.

(8.20)

Eh ben, dedans, mon je vas
dire dedans vingt-cinq ans, ou
trente ans d'icitte, il va plus
en rester qui va parler comme
nous-autres on parle cadien.

Well, in, I want to say in twenty-
five years, or thirty years from
now, there won't be any left who
can speak like we speak Cajun
[French].

Many were pessimistic regarding the possibility of revitalization of either the land or the language. One interviewee in Pointe aux Chênes was
particularly dismissive regarding the possibility of saving the eroding land:

(8.21)

L1: Et ça va pas menir meilleur.
R1: Ah, non?
L1: *Oh no.* Eusse peut tout faire
les huitres qu'eusse veut; c'est,
c'est *get* trop tard asteur.
R1: Hmm.
L1: *So* ça va faire un 'tit brin
du bien pour nous donner
le temps, si y a un mauvais
temps qui vient, pour nous-
autres s'en aller, mais . . .

L1: *And it's not going to get better.*
R1: *Oh, no?*
L1: *Oh, no. They can put down all*
the oysters they want; it's, it's
getting too late now.
R1: *Hmm.*
L1: *So it'll do some good, it'll give*
us a little time, if a hurricane
comes, to get out of the way,
but . . .

Another Pointe aux Chênes couple (see excerpt [8.19]) had an especially negative view of the prospect of language revival, dismissing it several times over the course of the conversation. In this case, Rocky, my field assistant, first brings up the importance of saving the language. Both interviewees agree but immediately issue dire predictions for its future ("you can say goodbye to it"). Rocky then asks what should be done to save the language and is met with dry laughter:

(8.22)

L1: {scoffing} Huh!
{laughter}
L1: Je crois pas tu peux sauver
    ça ici.

L1: *{scoffing} Huh!*
{laughter}
L1: *I don't think you can save this*
    here.

A few minutes later, she comments again,

(8.23)

L1: Ça fait, ça va s'en aller [tu]
    connais. Les enfants va pas,
    uh, pas montrer à eux-autres
    enfants c'e—c'est {hands slap}
    it's gone. {laughs}

L1: *So, it's going away, you know.*
    The kids aren't going to, uh,
    teach it to their kids, it's it's,
    {hands slap} it's gone. {laughs}

Her husband then chimes in with another timeline for the language's disappearance:

(8.24)

L2: Non, quand le vieux monde
    va—va être gone, je crois
    qu'on pourra oublier le
    français.

L2: *No, when the old people are—*
    are gone, I think you can
    forget about French.

Eventually, our motives are questioned: Are we trying to save the language? Rocky admits to making efforts to secure its future but is immediately shot down:

(8.25)

L1: T'aimerais sauver le français?	L1: *You'd like to save French?*
R2: Ouais.	R2: *Yeah.*
R1: Lui, il essaye.	R1: *He's trying.*
L1: Je, je crois pas que tu vas, tu vas *succeed*. {laughs}	L1: *I, I don't think you're going to, you're going to succeed {laughs}*
R3: Mais il faut essayer, hein?	R3: *But you gotta try, right?*
L1: Ouais faut que tu x essayes but ...	L1: *Yeah, you gotta try, but ...*

Though there is laughter throughout the conversation, that laughter is hollow, and the couple make it clear that they do not really see the decline of the language as a joke any more than others view the disappearance of the land as amusing. The laughter and jokes (e.g., we'll just get our feet wet) that punctuate these conversations should not be mistaken for insouciance. Rather, the tone was more along the lines of gallows humor.

A few people even directly compared the disappearance of the French language with that of the land.

"C'est tout après *gone*! [It's all disappearing]," shouted one interviewee from Leeville when I asked whether his children or grandchildren spoke French. His tone suggested a combination of scorn (at my question), frustration, and despair yet simultaneously and somewhat incongruously a tinge of humor. "Pareil comme la terre! [Just like the land!]," he added, sweeping his hand out to indicate the bayou and the sinking telephone poles behind us. "C'est tout après *gone*!"

The inevitable disappearance of the language is treated with the same fatalism yet resistance as is that of the land. The time depth of the problem and the attitude that speakers publicly espouse may suggest that the loss of both language and land are simply an accepted reality. This is true on some level but remains deeply upsetting to some residents. After all, Ignace's story of the destruction of Leeville was meant to both illustrate a long-standing state of affairs that underlies place-based identity on the bayou and serve as a dramatic warning to future generations. He predicted that in the near future, "Grand Island will be a painting of the past," comparing it to Îles Dernières, which had long been abandoned by the time he was born but whose warning was carried forward in the stories he grew up hearing. He told us that the people living in the coastal marshes were living on borrowed time: "I predict it a long time ago. You can say it, when it started, those levees, make those levees? Those levee that they dig. They still going on. Oh, Lord. Levees they hold the water back. That gonna be the end of those place, is those levee. All they need to break a levee."

He left off with a reference to the past as a warning to the future.

"The water's coming up, coming over, coming up," he said. "Just like Leeville."

CONCLUSION

Both Burley (2010) and Maldonado (2014) document a sense of fragility, a sense of solastalgia for a place that no longer exists as a result of the erosion of the land. The stories presented here suggest that this fragility is the result of the disappearance of a place both physically *and* linguistically. The invocation of French in telling the stories of past destruction, the overtly parallel ways in which the disappearance of the land and of the language are discussed, illustrate that aural and physical place are two sides of the same coin. It is consequently unsurprising that researchers in other fields studying attachment to place report that discussion of the disappearance of French arises with some frequency in their own studies of the area (Haertling 2019), and representatives of Louisiana's Strategic Adaptation for Future Environments (LASAFE) report that coastal residents discussing ways to mitigate the effects of coastal erosion and the potential need to move their communities to higher ground also constantly mention the need to maintain and/or revive the French language (Garner 2018).

The stories used to create place in Terrebonne-Lafourche highlight landmarks that are literally in the process of disappearing. If place is created by stories anchored by landmarks, a sense of solastalgia sets in when those landmarks are disappearing. The conclusion discusses the mechanisms by which aural place is constructed, showing the analogous structures that allow for a linguistic construction of place.

Chapter 9

CONCLUSION

In 2016, the US Department of Housing and Urban Development awarded $48.3 million to relocate the residents of the Île à Jean Charles to higher ground. This action caused news outlets to pronounce them the first American climate change refugees (e.g., Van Houten 2016; Schladebeck 2017). The idea of moving the community to higher ground was not new; I had discussed it in 2009 with Albert Naquin, principal chief of the Biloxi-Chitimacha Confederation of Muskogees, who had already been considering it for some time.[1] Our conversation highlighted the processes of place-making, both physical and linguistic, and the deep attachment to place described in this book.

Naquin explained that his family had occupied the island since at least 1840. Between 2002 and 2008, the community had been "drowned" (*noyé*) five times: by Hurricanes Lily (2002), Rita (2005), Gustav (2008), and Ike (2008) and by Tropical Storm Bill (2003). The community is outside the parish's levee protection area. As a result, houses had been destroyed beyond repair, and people had moved away from the island. Prior to 2002, the Island had been home to fifty-eight families, but by 2009, only twenty-five remained. In a separate interview, Father Roch Naquin, a retired priest native to the Island, estimated the number of current residents at roughly eighty.[2] Chief Naquin explained that his community was now spread out, with people (including him) living in Pointe aux Chênes as well as in Montegut, Petit Caillou, Houma, and beyond. Younger residents—people in their productive years—also frequently moved away in search of economic opportunity or because they had married outside the community. The remaining residents were primarily elderly people and the grandchildren they cared for while their children were at work. Chief Naquin dreamed of rebuilding his community as it had once been by gathering people back together somewhere safe.

"Ça pourrait nommer ça peut-être l'Île à Jean Charles numéro deux [They could maybe call it the Île à Jean Charles No. 2]," he said with a slight smile, clearly only half joking. "Comme ça nous donne la chance de recommencer notre vie. Ça pourrait nous ramener à *back* dans un petit pays [It would give us a chance to restart our lives. It could bring us all back together in a little country]."

He described the island in his youth: thick trees, including pecans and oranges, gardens growing abundant vegetables, grazing cows and some goats, community *boucheries* (butcherings of animals that were then shared with the group). He also described the recent measures taken to try to stop the rapid erosion of the land. Finally, he discussed the role French played in community life. When he was young, a few elderly residents had still spoken an indigenous language, but French was the language of everyday life. The generation above his, he said, had spoken better French than he did: for example, they said *fils* instead of *garçon* for "son." But most important, everywhere they went, their interactions—with neighbors, with shopkeepers, and so on—were in French. Today, when young people left the island, they left the language behind, too. The youth were no longer taught the language, but even middle-aged people like the chief who had grown up speaking it often gave it up in the face of peer pressure outside the lower bayous. Noting that his own children did not speak French, he pointed out, "En un generation, on a perdu le français [In one generation, we lost French]."

Chief Naquin felt that maintaining the language was very important but that the only way to do so was for it to be part of community life, as it had been when he was growing up:

(9.1)

Il faudrait grouiller dans un pays justement pour nous-autres, et les gens parlait français. On aurait comme un petit *community center*? Et tous les aprèsmidi autour de trois heures on se rencontrerait tous là, et boire du café, et parler de qui on a fait dans la journée. Eux-autres qui travaille pas, comme tous les plus vieux, et ça amènerait les petits avec eusse et parler français et peut-être qu'on pourrait le recommencer *back*.	*We would have to move to a country just for us, where everyone speaks French. We would have like a little community center? And every day about 3:00 we would meet there to drink coffee and to talk about what we'd done that day. The people who don't work, like all the old people. And they would bring the kids with them, and speak French and maybe we could bring it back.*

It wouldn't be the French the old people had spoken, he conceded, but it would be the French they used now, and that was good enough. He concluded, "*So* si on pourrait faire ça, ça c'est mon portrait que je vois dans [ma tête], ma vision [So, if we could do that, that's the picture I have in my head, my vision]."

Chief Naquin's dream to rebuild his community on high ground unites the themes developed throughout this book. First, he describes the ongoing destruction to the land and the way it was in the old days—the same kinds of stories we saw in chapter 8. Then he suggests that the community be relocated and that the new town should be named Île à Jean Charles No. 2. While he may only be half-serious about the name (*New Île à Jean Charles* would be a catchier and more conventional moniker, for starters), he is serious about claiming a space, about creating place via naming; the name he proposes merely stresses the fact that it would be the same community, the same *place*, gathering in this new space. The name characterizes the place, as did the name of Pointe au(x) Ch(i)en(es) in chapter 6. In this case, it stresses that the new place will have the same positive attributes as the old. He then details the disappearance of French, stressing that the language of the community should be French and specifically *their* French. He has to make some concessions as to its quality given the circumstances, but it retains enough of its character in its current form to suffice. The inclusion of French in Chief Naquin's vision is key to the construction of a new Île à Jean Charles.

Chapter 4 illustrates the strength of people's attachment to their local dialect. This attachment is not surprising: the local dialect is the repository of history; it carries the stories of the community. It goes beyond this, however: in the same way that to speak the name of a place is to recall the stories used in its creation, to use a language is to reference the stories it tells. If storytelling is place-making, then simply using the language is to create place in that the language invokes stories that created the place. But what are the specific mechanisms by which use of the language makes this possible?

Schafer (1993) defines the soundscape as the acoustic environment, including human voices, and identifies three categories of sound: keynote sounds, signals, and soundmarks. Keynotes exist in the background, setting the tone. Signals are sounds that are deliberately listened to; a keynote sound may become a signal if it is consciously identified. Soundmarks are unique sounds that therefore may identify a location. The process of enregisterment described by Johnstone, Andrus, and Danielson (2006) is analogous to this process of focusing on a keynote and creating from it a soundmark. Just as Johnstone (2013) shows for Pittsburgh, though in a very different set of circumstances, language is used to define community. The features outlined in chapter 4 that identify someone as a resident of the bayou—the features that delimit the boundaries of the dialect—are soundmarks. Though perhaps not limited to a given place, they are perceived to be so; these soundmarks are the aural equivalent of landmarks, the physical features that identify the boundaries of a place on the landscape. In this way, aural place is created in parallel ways to physical place.

Basso (1996b:7) tells us that what people make of their places they make of themselves. Place, as the marking of the boundaries encompassing a community, is inherently personal. This is why we see people arguing over the name of a town in chapter 6. The argument does not concern which group will get stewardship over the place, as is evident from the fact that they put less stock into the ethnic and subregional divisions in their community, as chapter 5 shows. Instead, there are factions in the town arguing over the characterization of the place: Are they the people from Dog Point or Oak Point? People's associations with oaks and dogs have potential repercussions on the identities of the people who live in the place. Silverstein (2014) describes this when he remarks on a growing national phenomenon of pride in what he terms *placèdness*. What were once first-order markers of geographic affiliation, he notes, have become second-order indicators of the positive values held by the people who live in a place.

That place is personal is reflected in the strength of the feeling of connection that people have for both the land and the language. The fierce attachment to the land, as Toot Naquin demonstrated with her declaration that she would never move and would die on the bayou on which she was raised, was echoed by other residents, both Cajun and Indian:

(9.2)

Je suis né et élevé icitte à l'île. Et je vas mourir ici si le Bon Dieu veut.	*I was born and raised here on the Island. And I'll die here, God willing.*
	(Indian, Île à Jean Charles)

(9.3)

Now, I'm not gonna move from here. I'll die here, I guess. I like it down here.
(Cajun, Pointe aux Chênes)

(9.4)

Ina plein plein que, qu'a parti.	*There are lots [of people] who*
Mon, je vas pas m'en aller. Je	*have left. Me, I'm not going.*
[suis] assez vieux, je peux finir	*I'm old enough, I can finish the*
la balance de ma vie ici là.	*rest of my life here. When we go*
Équand va débarquer en dehors	*outside, we'll wash our feet!*
s—on va se laver les pieds!	(Indian, Point aux Chênes)
{chuckles}	

The loss of the language also engenders strong sentiment, as chapter 8 shows. One interviewee on the Île à Jean Charles, however, characterized not having taught his children to speak French as not having raised them well— that is, he saw himself as having committed some sort of moral transgression:

(9.5)

L1: Mon, j'ai pas élevé mes enfants bien mon eusse peut pas—ça parle tout en anglais.	*L1: Me, I didn't raise my kids right, they can't—they all speak English.*
R1: Ah ouais.	*R1: Oh, really.*
L1: Et, je les ai pas montré bien du tout. Ina qui parle français des jeunes, en bas là après passer le petit pont.	*L1: And, I didn't teach them well at all. There are some young people who speak French, down there below the little bridge.*
R1: Uh-huh?	*R1: Uh-huh?*
L1: Eusse a élevé leurs enfants bien là mon j'ai pas élevé mes enfants bien du tout. Ça parle pas français du tout. Un petit peu, mais pas un tas.	*L1: They raised their kids right. Me, I didn't raise my kids right at all. They don't speak French at all. A little bit, but not a lot.*

Because both language and land are disappearing in Terrebonne-Lafourche, we can see the parallels in their construction of place; when land and language are robust, their associations with each other are readily and overtly apparent. But because both are disappearing at the same time, the losses are felt acutely and have similar impacts. In the only mention of French in his text, Burley (2010:48) cites the example of Vivian, a fifty-year-old educator native to Grand Isle who is seeking a new home on higher ground:

I'm moving to Thibodaux. I'm gonna live in Thibodaux, Louisiana. It's still French-speaking. They still have some French-speaking people there, and they're not going to be beachfront property for fifty years, so I'll be gone by then. I was gonna move to Lafayette, but it's too far away, I think. So Thibodaux.

It would be fair to argue that Thibodaux is in many ways already Vivian's home, since the connections between Grand Isle and Lafourche Parish are many—the closest medical services, restaurants, and major stores to Grand Isle are in lower Lafourche, and Grand Isle residents commonly travel as far as Houma or Thibodaux in search of services. But if language and land are both fundamental to identity, then losing one is traumatic enough; as Vivian demonstrates, losing both would be too much.

Place creation is fundamentally the mapping of a community's boundaries onto its space. This is done via storytelling, using landmarks to define the boundaries of physical space, and it is done via the use of soundmarks to indicate the boundaries of aural space. Places reflect who we are and teach us how to be. They are intrinsically personal, and their loss is felt profoundly. There are many reasons that we should be concerned about land and language loss: for example, the loss of protective wetlands leaves coastal communities more vulnerable to tidal surges and threatens the habitat of many species of wildlife, and the reduction of linguistic diversity through language death closes windows onto understanding human cognition. But equally important is compassion. It is precisely because the connection to land and to language is so very intimate that they matter so much to the people whose lives shape and are shaped by them and that we should care when they are threatened.

EPILOGUE

In February 2008, Toot fulfilled her promise. She had been sick for some time, and I received a call saying that the doctors had said that the end was near. I drove down to Pointe aux Chênes to say good-bye.

The house—especially Toot's room—was full of people. Her family and friends were sitting around her bed, playing CDs of her favorite songs, talking and laughing about old times and new, periodically addressing Toot, who remained unresponsive, though I was told that she was still capable of responding forcefully when she really wanted to—she had spat out her dentures in disgust when someone had put them in her mouth, for example. Prayer was another popular activity, and I was asked to look up the text of the Lord's Prayer in French on the Internet and to then lead the group in prayer. It was important to everyone present that the prayer be said in French. I left sometime in the late afternoon, with a large crowd remaining in the room behind me. Cheril, Toot's first cousin once removed as well as her goddaughter, recounted Toot's last minutes to me a few days later, after her funeral.

People had been coming in and out of the house for days, but when Cheril and her mother, Toot's cousin Alida, arrived at about 9:20 the next morning, it was quiet. Toot was wide awake for the first time in days. She stared at them. Cheril asked her a few questions, and while she was only able to give yes or no answers, it was clear she had been aware of what was going on around her over the past few days.

Speaking in French, Alida asked Toot if she wanted to pray together. Alida's voice is very quiet, so Cheril had to repeat the question. Toot said yes. So Cheril left the room, and Alida started praying in her low voice. She got through the Lord's Prayer, the Hail Mary, and two more, and then Toot passed. It seemed appropriate that Toot's last conversation was with Alida, in French, in the house she built on the land her father had bought high above the flood line.

NOTES

CHAPTER 1: SHIFTING LAND, SHIFTING LANGUAGE

1. Some residents prefer the name Pointe au Chien for both the bayou and the town; this has long been a matter of dispute. For the sake of simplicity, I use the names that appear most often on maps and in publications: Pointe au Chien for the bayou and Pointe aux Chênes for the town. I discuss the dispute and its implications in depth in chapter 6.

2. In Louisiana, secular parishes are equivalent to counties elsewhere in the country.

3. Among the problems with the census are respondents' shame at speaking nonstandard French and consequent reluctance to admit doing so and the possibility that they overcome that shame and answer differently on later censuses. In addition, the wording of the question was "Does this person speak a language other than English at home?" Many francophones, particularly those living with younger anglophone relatives, no longer speak French in the home and would consequently answer in the negative. Likewise, the number of people claiming to speak Creole, French, or Patois is also muddied by immigrants from places such as France or Haiti: the rise in the number of Creole speakers between 2000 and 2010 undoubtedly results at least partly from the fact that the number of Haitian immigrants (nearly all of whom live in the Greater New Orleans Metropolitan area) more than doubled between 2010 and 2017 (US Census Bureau 2010, 2017). Likewise, many residents are suspicious of the census or simply uninterested in filling it out, as I discovered one day while discussing the issues related to wording at a meeting in Pointe aux Chênes. "That's nothing," my amused interlocutor told me. "Watch this."

She turned to a group of elderly francophones sitting along the wall.

"Hey," she addressed them, "What do y'all do with the census forms when you get them?"

The group replied in unison, "We throw them away!"

4. Many publications—particularly newspapers—call the island the Île *de* Jean Charles. Throughout this book, I use the name used by its residents and their neighbors, which follows the grammar of Louisiana Regional French and uses the preposition *à* rather than *de* to indicate possession. The most frequent pronunciation I hear in English leaves out the preposition altogether.

5. My home country, Canada, has long rejected the term *Indian* in favor of *First Nations* (or *Métis*, depending on the context), and it also provokes controversy in the United States. Despite its potential problems, however, community members with whom I worked

soundly rejected *First Nations* and *Native American* and preferred *Indian*. I therefore use the term *Indian* here with caution but out of respect for their stated preferences.

6. Literally, "No hurricane is going to take me!"

7. Among the local rituals I attended were the annual blessing of the shrimp boat fleet, the French Mass at Live Oak Baptist Church in lower Pointe aux Chênes, Halloween events at the local Pentecostal church, Christmas boat parades and bonfires, Carnival parties and parades in Golden Meadow, and the Indian Mass at St. Charles Borromeo Catholic Church in upper Pointe aux Chênes.

CHAPTER 2: THE LAND AND ITS PEOPLE: DESCRIPTION OF THE LOWER LAFOURCHE BASIN

1. In this book, I use the term *lower Lafourche Basin* to indicate the portion of the Lafourche delta encompassed by Terrebonne and Lafourche Parishes. This is not to be confused (though one can easily see how it might be) with *lower Lafourche Parish*, which I use to indicate that part of Lafourche Parish that lies below the Intracoastal Canal.

2. The term *densely* should be qualified. A resident of Paris or New York—or even New Orleans—would find the suggestion that the area is densely populated laughably inaccurate, given the single-story dwellings and the overall population of less than twenty-five thousand for the roughly sixteen-mile corridor running between Larose and Golden Meadow. The bayou is, however, *relatively* densely populated in that it is continuously occupied from Donaldsonville through Golden Meadow; the lower bayou below the Intracoastal Canal is flanked by a near-constant parade of single-story dwellings, and small residential streets several hundred feet deep branch off the main thoroughfares of Louisiana Highway 1 and Louisiana Highway 308 in each of the towns, and the boundaries between them are not clearly demarcated.

3. The density of the Indian population of Golden Meadow may historically have been higher: today, Indians have spread from the Indian Settlement below Golden Meadow to Larose (and occasionally to Houma; in fact, a move to Houma is common especially of younger generations across all the bayous). The percentage of Indians in the population on lower Bayou Lafourche per the 2000 US Census is as follows: Galliano—4.5 percent, Cut Off—3.8 percent, Larose—3.9 percent.

4. That is to say, two versions not including the one put forth by Indian agents reporting in the early nineteenth century that has been at the root of the struggle for federal recognition. That version asserts that the Houma disappeared but does not address whether their disappearance resulted from death, assimilation into the surrounding population, or both.

5. Unless otherwise noted, the information on the movement of Indian groups during this period is taken from Westerman n.d.

6. A full discussion of the findings is too complex to discuss here. However, the findings differed from those of the UHN in that the largest problem for these groups was a gap in documented leadership between 1940 and 1980 (US Bureau of Indian Affairs 2008a, b).

7. The figure is based on a count including people who claimed Black or African American alone or combination with other races.

8. Of course, some of these people were likely also free before the war, and many of the enslaved were of mixed race and often passed for white—and did so on the census. Still, for want of a better starting point, the assumption that most of those listed as black are former slaves or their children is a working solution.

9. Records housed at the Terrebonne Parish Library, collections as seen at https://mytpl. org/genealogy/.

10. It is also possible, as Picone (2003) suggests, that some francophones deliberately sought out anglophone slaves to teach the slaveholders' children English.

CHAPTER 3: THE HISTORY OF FRENCH IN LOUISIANA

1. In the case of other languages, that influence is in all likelihood limited to lexical borrowing, though Faraclas (2010) makes the intriguing suggestion that indigenous languages have affected it prosodically for indigenous speakers.

2. While Spain did own the colony and a small number of Spanish-speaking immigrants did settle in Louisiana, the majority and therefore hegemonic language during this period remained French.

3. As a side note, other labels used include *French* or *broken French* (for any variety), *Kouri-vini, gumbo, français nègre,* and others (for Louisiana Creole). A number of people prefer to use *Louisiana French*; the term is used by some to refer only to Louisiana Regional French and by others as a general label that permits inclusiveness when considering Louisiana's francophone scene, for example. In these cases it refers to Louisiana Regional French and Louisiana Creole together. Due to its ambiguity in this regard, the term *Louisiana Regional French* is preferable in the context of an academic study. In any case, *French* is the term most commonly used by speakers, with people choosing another term, most often *Cajun French* or *Creole,* when called on to be more specific.

4. Similar shorthanding can be found in many publications including, ironically and by admission of the author himself, in work by Klingler (2009), who has gone to great lengths to debunk the myth.

5. "En fin de compte, ce n'est qu'au sein de la population d'origine acadienne isolée . . . que se conserva une variété de français, le FC" (Picone and Valdman 1995:150).

6. Lyche (1995) provides an alternate analysis for this phenomenon, suggesting that these forms can be accounted for by epenthesis of an element following the deletion of schwa (*cette* is first altered to [stə], a common phenomenon across French dialects).

7. For more complete accounts of this history, see Brasseaux 2005; Klingler 2009.

8. The information in this section comes from Brasseaux (1987, 1992, 2005) and Klingler (2009).

9. All demographic data on early settlement is taken from Brasseaux 2005.

10. There were, of course, many tens of thousands of indigenous people living in what is now Louisiana, and many of them had daily interactions with the colonists. For various reasons their languages did not become the spoken standard of the colony, though Mobilian Jargon, a trade language based on Choctaw, was widely spoken and used in interactions with colonists.

CHAPTER 4: FRENCH IN THE LOWER LAFOURCHE BASIN

1. When we presented our results at a public forum that included several of our interviewees, they said that we had "made too much of" the differences. Presumably we had given the impression that the variation we found was enough to warrant the easy identification of two different varieties of French, something that was not at all our intent.

2. This methodology is not without its limitations. For example, some speakers may be motivated instead to perform their most nonstandard for the interviewer in such a context. Still, the results I received, in which "correct" speech parallels standard usage in modern France, suggest it was generally successful.

3. Trudgill (1986) describes a very similar case in Norwich, England, involving the vowel in such words as *room*, in which the variants index social class. That phenomenon is absent in the context described here.

4. The change in tense is common in such exercises and does not affect the choice of pronoun.

5. The Île à Jean Charles lacks a comparative Cajun figure simply because I could find no one there who claimed to be exclusively Cajun—everyone claimed both heritages but stressed the Indian component.

6. Faraclas (2010) claims that the seventy-five words from Swanton's list are still used by the community, but I have only encountered confusion regarding this claim. This confusion was expressed during a conversation I was witness to but not a part of, and while it is possible that a show was being put on for me, that seems unlikely, and consequently I have no reason to believe they were not telling the truth.

7. A deceased resident of the Île à Jean Charles whose house still stood was named "Chocolat" [ʃokola]; his name may have in fact been [ʧakalata], i.e., Hawk.

8. That /y/ is a sound found in French but not any of the languages spoken by the indigenous groups the community lists as ancestors suggests that [kyl] is either a word of French origin or modified by French. I have found interesting similarities between [ʧakalata] and various terms in Ishak (Atakapa): a bird, for example, is a *ts'ok*, and a buffalo is a *shok*. Given that residents of the Island were nicknamed *Buffalos*, the possibility that Chocolat's nickname was in fact a play on *buffalo* in some way is interesting, but the similarities are vague and would require compounding, which as a nonexpert in these languages I may well be doing incorrectly. Consequently, I hesitate to make any strong statements in support of any such link.

9. Indeed, at a Tunica-Biloxi powwow in Marksville in 2012, I attended a presentation on Tunica songs that featured a French song; the presenters explained to the crowd that French was as important to the community as Tunica.

CHAPTER 6: OAK POINT OR DOG POINT? THE IMPORTANCE OF A NAME

1. The missing circumflex accent is not accidental. It is in fact absent from most documents; this absence is in most cases likely not indicative of anything but an inability to produce a circumflex accent with typesetting/word processors. In any case, in this discussion,

all cited names appear as they do in the original. Readers may assume that if an accent is present in this text, it was present in the original, and vice versa. Likewise, differences in capitalization and the variation in the spellings of *Point(e)* and *au(x)* and the final *s* on *chênes*.

2. Though Read ([1931] 1963:178) cites "a survey of 1856" as recording the name including *Pointe*. As a point of potential interest, a certain *River aux Chenes* in nearby Plaquemines Parish seems to have suffered the same identity issues, as the GNIS lists six name variants, three of which are variants of *Chenes* (US Geological Survey 2019b); the other three are Riviere aux Chiens (twice) and Dog River, the latter a direct translation of the former and bearing the same citation. The original source of the *Chien* (Dog) variant in this case appears to be Andre Penicaut, the French colonist who accompanied Iberville in 1700, lived among the Natchez, and wrote a memoir of his travels.

3. The minutes of the Terrebonne Parish Police Jury forwarded to me by the current Terrebonne Parish Council give no date beyond the year 1978 and record that the parish is moving that "the proper Resolution be forwarded to Terrebonne Parish's legislative Delegation, Louisiana Department of Public Works, the Governor and Louisiana's Congressional Delegation requesting that the name be changed and that any future maps be corrected relative to the name of the community of Pointe Aux Chenes; and that this matter be followed up by any correspondence necessary; and that a copy of this Resolution be forwarded to the Louisiana Highway Department, the Waterworks District, and the Department of Wildlife and Fisheries." Given that the Wildlife Management Area was renamed independently in 2001 via a concurrent resolution of the legislature, it is unclear that the name change was finalized at the police jury stage. In any case, current employees of the state legislature were unable to find any items relevant to this in their archives, so I accept here the version printed in the *Baton Rouge Advocate* (Associated Press 1986) that attributes the name change to the Police Jury, though Ledet's relatives claim, and the minutes of the police jury meeting seem to corroborate, that the matter was forwarded to the legislature. Whether the legislature did—or needed to do—anything after that is beyond my ability to discern.

4. Though I have frequently also heard an ambiguous *Chein*—i.e., [ʃɛ̃], lacking both the glide and the final /n/—from English speakers; French speakers wishing to avoid the issue say *Chiennes* [ʃjɛn].

5. The trees and the dog seem to be a play on the town's dual names, though it isn't clear that this joke was intentional. It was not discussed at meetings discussing the design of the sign, and it is equally possible that the image of large, healthy trees was simply meant to celebrate or commemorate the health of the land in the past in stark contrast to the increasingly thin forest and dead trees found today and that the visual pun is accidental. Nobody argues that there were no oaks on the Point; they argue only that it was not named for them.

6. Though the fish could go either way. In this case, however, it appears to be an image of sports fishing (given the line and the fish in the air) rather than subsistence fishing.

7. Both searches also include results that include the search term as part of the name—for example, *Oakley* or, more problematically, *Oakwuroot* or *Dogie*.

8. That name is *Sabine*. I hesitated to include it here given the deep hurt that may result from its use, even when not directed toward someone. However, given that existing (older) literature often uses the term neutrally, I feel it important to make it clear to readers who might encounter it in such a context elsewhere that the term is in fact very pejorative and should be avoided at all cost.

CHAPTER 7: LANGUAGE SHIFT AND THE CONTINUED IMPORTANCE OF FRENCH TO BAYOU IDENTITY

1. When survey participants were asked how they identified ethnically, a few wrote in the controversial term *coonass* rather than Cajun. This identity phenomenon merits further study but is beyond the scope of this volume.

2. A "groupe rythmique" is a phrase, generally corresponding to a syntactic phrase, though grouping can change based on such things as rate of speech and stylistic framing of information units and therefore may defy easy definition. In any case, a single utterance may contain more than one "groupe rythmique" and therefore may have more than one syllable that bears primary stress.

3. Despite the presence of pejoratives like *redneck* and *hillbilly*, absent from the maps is the term *coonass*. Though nobody labeled the maps this way, two participants (both white males under the age of forty) chose to write in the term as an ethnic identifier. Perhaps tellingly, for one of those respondents, it was fairly clear that the lines referred to French rather than English variation.

CHAPTER 8: OMENS FROM THE PAST, WARNINGS FOR THE FUTURE: STORYTELLING, PLACE, AND IDENTITY

1. The quoted text presented in this chapter is an abridged version of my transcript of the interview with Ignace. I have deleted false starts and hesitations, and I have regularized the punctuation. In some cases, the story was told in fits and spurts, and rather than transcribe only a line at a time, I have summarized those portions in prose while trying to keep his story as intact as possible. He was a master storyteller, and eager to have his story heard. I have left the morphology and syntax as it is in the original to capture as much as possible of Ignace's speech, which was expressive and an integral part of the story's telling,

2. A *seine* is a French word for a net. *Seining*, then, means fishing with a net.

3. A common exclamation used to indicate surprise, appreciation, or, in this case, scorn.

4. Like a *seine*, a *tramail* is a type of fishing net; the *Dictionary of Louisiana French* (Valdman et al. 2010) attests both in addition to *paupière*, *épervier*, and *carrelet*. Ignace specified that a *seine* was for fish and a *tramail* for shrimp.

5. A clipped form of *petit* "small." Pronounced [ti]. It is frequently used in South Louisiana nicknames, especially for men. Many prefer to spell it *Tee* or *T-*, e.g., *T-Boy*.

6. Oyster shells are used to create artificial reefs to stop or slow erosion.

CHAPTER 9: CONCLUSION

1. The effort to relocate the community failed in 2019 when Naquin withdrew his support for the project, citing leaders' unwillingness to respect the local culture.

2. Lagamayo 2017 estimates thirty families, while Van Houten 2016 cites one resident as estimating sixty residents and Hermann 2017 gives the population as around seventy.

WORKS CITED

Abney, Lisa. 2019. "The Linguistic Survey of North Louisiana: History and Progress." In *Language in Louisiana: Community and Culture*, edited by Nathalie Dajko and Shana Walton, 203–26. Jackson: University Press of Mississippi.

Agha, Asif. 2003. "The Social Life of Cultural Value." *Language and Communication* 23:231–73.

AJ+. 2015. "Louisiana Disappearing: Living on the Brink of Climate Change." Video recording. Hosted by Francesca Fiorentini. Produced by AJ+. https://vid.wxzm.sx/watch?v=4THdX9KOZ_4.

Allen, Barbara. 1986. "Numa Boudreaux." *Terrebonne Life Lines* 5 (4):19–28.

Alvarez, Louis, and Andrew Kolker. 1988. *American Tongues*. Video recording. Produced and directed by Louis Alvarez and Andrew Kolker. New York: Center for New American Media.

American Sugar Cane League. 2018. "Raw Sugar Factories." *American Sugar Cane League: Making Life Sweeter. Naturally.* https://www.amscl.org/industry-info/raw-sugar-factories/.

Associated Press. 1986. "Community's Name Changed." *Baton Rouge Morning Advocate*, December 23, 4-B.

Ayres-Bennet, Wendy. 2004. *Sociolinguistic Variation in Seventeenth-Century France*. Cambridge: Cambridge University Press.

Barrow, Robert H. 1854. "Ranaway from the Subscriber, the Negro Man Tom Anderson." *American Patriot*, December 27, 3. *Louisiana Runaway Slave Advertisements, 1836–1865*, http://www.louisianadigitallibrary.org/islandora/object/lsu-p16313coll80%3A4036.

Barrow, Robert Ruffin, Papers. #2407-z, Southern Historical Collection, Wilson Library, University of North Carolina, Chapel Hill.

Barrow, Robert Ruffin. Plantation Journal, 1833–35. Barrow Family Papers. Manuscripts Collection 654, Howard Tilton Memorial Library, Tulane University, New Orleans.

Basso, Keith. 1996a. "Wisdom Sits in Places." In *Senses of Place*, edited by Steven Feld and Keith Basso, 53–90. Santa Fe: School of American Research Press.

Basso, Keith. 1996b. *Wisdom Sits in Places: Landscape and Language among the Western Apache*. Albuquerque: University of New Mexico Press.

Bayley, G. W. R. 1853. "New and Improved Map of Louisiana." Louisiana Research Collection, Tulane University.

Bernard, Shane K. 2003. *The Cajuns: Americanization of a People*. Jackson: University Press of Mississippi.

Blu, Karen. 1996. "Where Do You Stay At? Homeplace and Community among the Lumbee." In *Senses of Place*, edited by Steven Feld and Keith Basso, 197–228. Santa Fe: School of American Research Press.

Bodin, Catherine. 1987. "The Dialectal Origins of Louisiana Acadian French." PhD diss., University of North Carolina–Chapel Hill.

Bocage, Charles William. 1916. *Bocage's Official Map of the Parish of Terrebonne, Louisiana.* New Orleans: William Bocage. https://www.loc.gov/item/2013593065.

Boissonneault, Chantal. 1999. "Le français de l'Abitibi: Caractéristiques phonétiques et origine socio-géographique des locuteurs." Master's thesis, Université Laval.

Borgatti, Stephen P. 1996. *Anthropac 4.0.* Natick, MA: Analytic Technologies.

Bouchereau, Alcee. 1878–1918. *Statement of the Sugar and Rice Crops Made in Louisiana.* New Orleans: Young, Bright. Available at Documenting Louisiana Sugar, http://www.sussex .ac.uk/louisianasugar/1-6.html.

Bouchereau, Louis. 1869–78. *Statement of the Sugar and Rice Crops Made in Louisiana.* New Orleans: Young, Bright. Available at Documenting Louisiana Sugar, http://www.sussex .ac.uk/louisianasugar/1-6.html.

Bourgeois, Henry L. 1938. "Four Decades of Public Education in Terrebonne Parish." Master's thesis, Louisiana State University.

Bowman, Greg, and Janel Curry-Roper. 1982. *The Houma People of Louisiana: A Story of Indian Survival.* Houma: United Houma Nation.

Brasseaux, Carl A. 1985. "Acadian Life in the Lafourche Country, 1766–1803." In *The Lafourche Country: The People and the Land,* edited by Phillip Uzee, 33–43. Thibodaux: Lafourche Heritage Society.

Brasseaux, Carl A. 1987. *The Founding of New Acadia.* Baton Rouge: Louisiana State University Press.

Brasseaux, Carl A. 1992. *Acadian to Cajun: Transformation of a People, 1803–1877.* Jackson: University Press of Mississippi.

Brasseaux, Carl A. 1996. "Acadian Life in the Lafourche Basin, 1803–1860." In *The Lafourche Country II: The Heritage and Its Keepers,* edited by Stephen S. Michot and John P. Doucet, 21–25. Thibodaux: Lafourche Heritage Society.

Brasseaux, Carl A. 2005. *French, Cajun, Creole, Houma: A Primer on Francophone Louisiana.* Baton Rouge: Louisiana State University Press.

Brown, Cecil H., and Heather K. Hardy. 2000. "What Is Houma?" *International Journal of American Linguistics* 66 (4): 521–48.

Brundage, W. Fitzhugh. 2000. "Le Reveil de la Louisiane: Memory and Acadian identity, 1920–1960." In *Where These Memories Grow: History, Memory, and Southern Identity,* edited by W. Fitzhugh Brundage, 271–98. Chapel Hill: University of North Carolina Press.

Burley, David. 2010. *Losing Ground: Identity and Land Loss in Coastal Louisiana.* Jackson: University Press of Mississippi.

Caldas, Stephen J. 2007. "French in Louisiana: A View from the Ground." In *French Applied Linguistics,* edited by Dalila Ayoun, 450–77. Amsterdam: Benjamins.

Calder v. Police Jury of Terrebonne. 1892. 44 La.Ann. 173. Supreme Court of Louisiana. No. 10,963. February 8.

Calvet, Louis-Jean. 1990. "Des mots sur les murs: Le marquage linguistique du territoire." *Migrants-Formation* 83:149–60. http://www2.cndp.fr/revueVEI/83/MigFo83-13.htm.

Carmichael, Katie. 2007. "The Substitution of /h/ for [ʒ] in Louisiana French, and Its Relation to Register." Honors thesis, Tulane University.

Chamberlin, J. Edward. 2001. "From Hand to Mouth: The Postcolonial Politics of Oral and Written Traditions." In *Reclaiming Indigenous Voice and Vision*, edited by Marie Battiste, 124–41. Vancouver: University of British Columbia Press.

Chambers, J. K., and Peter Trudgill. 1998. *Dialectology.* Cambridge: Cambridge University Press.

Champomier, P. A. 1846–62. *Statement of the Sugar and Rice Crops Made in Louisiana.* New Orleans: Cook, Young. Available at Documenting Louisiana Sugar, http://www.sussex.ac.uk/louisianasugar/index.php.

Charbonneau, René. 1957. "La spirantisation du /Z/." *Journal of the Canadian Linguistics Association/Revue de l'Association canadienne de linguistique* 3:14–19.

Cheramie, Deany M., and Donald A. Gill. 1992. "Lexical Choice in Cajun Vernacular English." In *Cajun Vernacular English: Informal English in French Louisiana*, edited by Ann Martin Scott, 38–55. Lafayette: University of Southwestern Louisiana Press.

Christie, Douglas E. 2009. "Place-Making as Contemplative Practice." *Anglican Theological Review* 91 (3): 347–72.

Coastal Protection and Restoration Authority of Louisiana. 2017. *Louisiana's Comprehensive Master Plan for a Sustainable Coast.* Baton Rouge: Coastal Protection and Restoration Authority of Louisiana. http://coastal.la.gov/wp-content/uploads/2017/04/2017-Coastal-Master-Plan_Web-Single-Page_CFinal-with-Effective-Date-06092017.pdf.

Concordia, with Chicago Bridge and Iron Company (CB&I), directed by Pan American Engineers (PAE). 2016. "The Resettlement of Isle de Jean Charles: Report on Data Gathering and Engagement Phase." Report delivered to the State of Louisiana Division of Administration, Office of Community Development, Disaster Recovery Unit (OCD-DRU), November 28 (updated May 2017).

Connor, Linda, Glenn Albrecht, Nick Higginbotham, Sonia Freeman, and Wayne Smith. 2004. "Environmental Change and Human Health in Upper Hunter Communities of New South Wales, Australia." *EcoHealth* 1 (suppl. 2): 47–58.

Cox, Juanita. 1992. "A Study of the Linguistic Features of Cajun English." ERIC doc. no. ED352840. http://files.eric.L1.gov/fulltext/ED352840.pdf.

Cran, William, and Robert MacNeil. 2005. *Do You Speak American?* Video recording. Directed by William Cran and hosted by Robert MacNeil. New York: Thirteen/WNET.

Curtis, Joseph. 2003. "Letter to the Editor." *Reno Gazette-Journal*, December 11, 6A.

Dajko, Nathalie. 2009. "Ethnic and Geographic Variation in the French of the Lafourche Basin." PhD diss., Tulane University.

Dajko, Nathalie. 2018. "The Continuing Symbolic Importance of French in Louisiana." In *Language Variation in the New South: Contemporary Perspectives on Change and Variation*, edited by Jeffrey Reaser, Eric Wilbanks, Karissa Wojcik, and Walt Wolfram, 153–74. Chapel Hill: University of North Carolina Press.

Dajko, Nathalie, Zachary Hebert, Keith Bedney, Rebecca Beyer, Jonathan Garen, Audrey Gilmore, Molly Heiligman, Shane Lief, Sean Simonson, and Adebusola Adebesin. 2012. "Authenticity and Change in New Orleans English." *Southern Journal of Linguistics* 36 (1): 139–57.

Dardar, T. Mayheart. 2005. *Pointe Ouiski: The Houma People of LaFourche-Terrebonne.* New York: Jewish Fund for Justice.

Davenport, Coral, and Campbell Robertson. 2016. "Resettling the First American 'Climate Refugees.'" *New York Times,* May 3.

DeSantis, John. 2016. "Slave 'Reparations' to Include Terrebonne, Lafourche Kin." *Times of Houma/Thibodaux,* September 6. https://www.houmatimes.com/news/slave-reparations -to-include-terrebonne-lafourche-kin/article_f9160898-7470-11e6-82e0-5b53bc45bb9d .html.

Doucet, Carol James. 1970. "The Acadian French of Lafayette, Louisiana." Master's thesis, Louisiana State University.

Dubois, Sylvie, and Barbara M. Horvath. 1998a. "From Accent to Marker in Cajun English: A Study of Dialect Formation in Progress." *English World-Wide* 19 (2): 161–88.

Dubois, Sylvie, and Barbara M. Horvath. 1998b. "Let's Tink about Dat: Interdental Fricatives in Cajun English." *Language Variation and Change* 10 (3): 245–61.

Dubois, Sylvie, and Barbara M. Horvath. 2000. "When the Music Changes, You Change Too: Gender and Language Change in Cajun English." *Language Variation and Change* 11 (3): 287–313.

Dubois, Sylvie, and Barbara M. Horvath. 2002. "Sounding Cajun: The Rhetorical Use of Dialect in Speech and Writing." *American Speech* 77 (3): 264–87.

Dubois, Sylvie, and Barbara M. Horvath. 2003a. "Creoles and Cajuns: A Portrait in Black and White." *American Speech* 78 (2): 192–207.

Dubois, Sylvie, and Barbara M. Horvath. 2003b. "The English Vernacular of the Creoles of Louisiana." *Language Variation and Change* 15 (3): 253–86.

Dubois, Sylvie, and Barbara M. Horvath. 2003c. "Verbal Morphology in Cajun Vernacular English: A Comparison with Other Varieties of Southern English." *Journal of English Linguistics* 31 (1): 34–59.

Duchmann, Holly. 2017. "Slave's Photos Put Face to Tragic Reality." *Houma Courier,* March 29. http://www.houmatoday.com/news/20170329/slaves-photos-put-face-to-tragic-reality.

Dupre, Reggie. 2001. Senate Concurrent Resolution no. 35. http://www.legis.la.gov/legis/ BillSearch.aspx.

Failevic, Maurice. 1969. "Les Enfants du Francien: La Louisiane." *XXème Siècle: Portrait de la Francophonie,* produced by Igor Barrère and Pierre Desgraupes, June 24. Office national de radiodiffusion télévision française. Available at https://www.ina.fr/video/CAF93022180/ portrait-de-la-francophonie-video.html

Ferguson, Patty. N.d. "History/Background: Our Community Yesterday, Today, Tomorrow . . ." http://pactribe.tripod.com/id2.html.

Fischer, Ann. 1970. "History and Current Status of the Houma Indians." In *The American Indian Today,* edited by Stuart Levine and Nancy Oestreich Lurie, 212–35. Baltimore: Penguin.

Fleming, R. Lee. 2013. "List of Petitioners by State (as of November 12, 2013)." https://www .bia.gov/sites/bia.gov/files/assets/as-ia/ofa/admindocs/ListPetByState_2013-11-12.pdf .

Flikeid, Karin. 1984. *La Variation phonétique dans le parler acadien du nord-est du Nouveau Brunswick : Étude sociolinguistique.* New York: Lang.

Flikeid, Karin. 2005. "Structural Aspects and Current Sociolinguistic Situation of Acadian French." In *French and Creole in Louisiana,* edited by Albert Valdman, 255–86. New York: Plenum.

Foret, Michael J. 1996. "Native Americans of the Lafourche Country, 1543–1996." In *The Lafourche Country II: The Heritage and Its Keepers,* edited by Stephen S. Michot and John P. Doucet, 13–20. Thibodaux: Lafourche Heritage Society.

Fortier, Alcée. 1884–85. "The French Language in Louisiana and the Negro-French Dialect." *Transactions of the Modern Language Association of America* 1:96–111.

Fortier, Alcee. 1891. "The Acadians of Louisiana and Their Dialect." *Publications of the Modern Language Association of America* 6 (1): 64–94.

Fridland, Valerie. 2008. "Regional Differences in Perceiving Vowel Tokens on Southerness, Education, and Pleasantness Ratings." *Language Variation and Change* 20:67–83.

Geophysical Fluid Dynamics Laboratory. 2018. "Global Warming and Hurricanes: An Overview of Current Research Results." https://www.gfdl.noaa.gov/global-warming-and-hurricanes/.

Google Maps. 2017. "Pointe au Chien vs. Pointe aux Chênes." Accessed June 19. https://www.google.fr/maps/place/Point-Aux-Chenes,+Montegut,+LA+70377,+USA/@29.4751286,-90.499314,18z/data=!4m5!3m4!1s0x8620fe9008d9aabd:0xf35e256b1bd47a5b!8m2!3d29.4977177!4d-90.5534195?hl=en.

Greider, Thomas, and Lorraine Garkovich. 1994. "Landscapes: The Social Construction of Nature and the Environment." *Rural Sociology* 59 (1): 1–24.

Gruenewald, David A. 2003. "Foundations of Place: A Multidisciplinary Framework for Place-Conscious Education." *American Educational Research Journal* 40 (3): 619–54.

Guilbeau, John. 1936. "A Glossary of Variants from Standard French in La Fourche Parish." Master's thesis, Louisiana State University.

Guilbeau, John. 1950. "The French Spoken in Lafourche Parish, Louisiana." PhD diss., University of North Carolina–Chapel Hill.

Hardee, T. S. 1870. *Map Illustrating the Topography of New Orleans and of the Coast of Louisiana and Mississippi.* Louisiana Research Collection, Tulane University.

Hardee, T. S. 1871. *Hardee's Geographical, Historical, and Official Map of Mississippi.* New Orleans: Lewis, 1871. Louisiana Research Collection, Tulane University.

Hauer, Matthew E. 2017. "Migration Induced by Sea-Level Rise Could Reshape the US Population Landscape." *Nature Climate Change* 7:321–25.

Hauer, Matthew E., Jason M. Evans, and Deepak R. Mishra. 2016. "Millions Projected to Be at Risk from Sea-Level Rise in the Continental United States." *Nature Climate Change* 6:691–98.

Heitmann, John A. 1996. "The Nineteenth and Early Twentieth Century Sugar Industry." In *The Lafourche Country II: The Heritage and Its Keepers,* edited by Stepen S. Michot and John P. Doucet, 96–102. Thibodaux: Lafourche Heritage Society.

Hendry, Barbara. 2006. "The Power of Names: Place-Making and People-Making in the Riojan Wine Region." *Names* 54 (1): 23–54.

Hermann, Victoria. 2017. "Disappearing Landscapes in Louisiana: When Google Maps Can't Catch Up." *National Geographic Blog,* March 20. https://blog.nationalgeographic.org/2017/03/20/disappearing-landscapes-in-louisiana-when-google-maps-cant-catch-up/.

Hewson, John. 2000. *The French Language in Canada.* Munich: Lincom Europa.

Hicks, Edgar. N.d. "Index for They Came They Stayed." http://www.rootsweb.ancestry.com/~la terreb/theycame/index-1.htm.

Horiot, Brigitte, and Pierre Gauthier. 1995. "Les parlers du sud-ouest." In *Français de France et français du Canada: Les parlers de l'ouest de la France, du Québec et de l'Acadie,* edited by Pierre Gauthier and Thomas Lavois, 187–249. Lyon: Université Lyon III Jean Moulin, Centre d'Études Linguistiques Jacques Goudet.

Iannàccaro, Gabriele, and Vittorio Dell'Aquila. 2001. "Mapping Languages from Inside: Notes on Perceptual Dialectology." *Social and Cultural Geography* 2 (3): 265–80.

Ingold, Tim. 2008. "Bindings against Boundaries: Entanglements of Life in an Open World." *Environment and Planning* 40:1796–1810.

Inscoe, John. 2006. "Slave Names." In *The Encyclopedia of North Carolina,* edited by William S. Powell. Chapel Hill: University of North Carolina Press. https://www.ncpedia.org/slave-names.

Irvine, Judith, and Susan Gal. 2000. "Language Ideology and Linguistic Differentiation." In *Regimes of Language: Ideologies, Polities, and Identities,* edited by Paul Kroskrity, 35–84. Santa Fe: School of American Research Press.

Jacob, Ursin. 1859. "Detained in the Jail of St. John the Baptist Parish." *Le Meschacébé,* January 29, 2. Available at Louisiana Runaway Slave Advertisements, 1836–1865, http://www.louisianadigitallibrary.org/islandora/object/lsu-p16313coll80%3A4036.

Jarman, Neil. 1993. "Intersecting Belfast." In *Landscape, Politics and Perspectives,* edited by Barbara Bender, 107–39. Oxford: Berg.

Jennings, Ken. 2013. "Yes, There Is a Town Named Dildo: Ken Jennings Reports." CNN, December 10. https://www.cntraveler.com/stories/2013-12-10/maphead-dildo-newfoundland-canada.

Johnson, Alvin Jewett. 1866. *Johnson's Arkansas, Mississippi, and Louisiana.* New York: Johnson. https://www.davidrumsey.com/luna/servlet/detail/RUMSEY~8~1~305431~9 0075790:Johnson-s-Arkansas,-Mississippi,-An?sort=pub_list_no_initialsort%2Cpub_ date%2Cpub_list_no%2Cseries_no&qvq=w4s:/where%2FUnited%2BStates%2FLo uisiana;sort:pub_list_no_initialsort%2Cpub_date%2Cpub_list_no%2Cseries_no;lc: RUMSEY~8~1&mi=30&trs=40.

Johnstone, Barbara. 2013. *Speaking Pittsburghese: The Story of a Dialect.* New York: Oxford University Press.

Johnstone, Barbara, Jennifer Andrus, and Andrew Danielson. 2006. "Mobility, Indexicality, and the Enregisterment of 'Pittsburghese.'" *Journal of English Linguistics* 34 (2): 77–104.

Kennedy, Merrit. 2017. "Louisiana's Governor Declares State of Emergency over Disappearing Coastline." NPR, April 20. https://www.npr.org/sections/thetwo-way/2017/04/20/524896256/louisianas-governor-declares-state-of-emergency-over-disappearing-coastline.

Klingler, Thomas A. 2003a. *If I Could Turn My Tongue Like That: The Creole of Pointe Coupee Parish, Louisiana.* Baton Rouge: Louisiana State University Press.

Klingler, Thomas A. 2003b "Language Labels and Language Use among Cajuns and Creoles in Louisiana." *University of Pennsylvania Working Papers* in Linguistics 9 (2). Available at https://repository.upenn.edu/pwpl/vol9/iss2/8.

Klingler, Thomas A. 2009. "How Much Acadian Is There in Cajun?" In *Acadians and Cajuns: The Politics and Culture of French Minorities in North America,* edited by Ursula Mathis-Mosen and Günter Beschof, 91–103. Innsbruck: Innsbruck University Press.

Klingler, Thomas A. 2014. "Variation phonétique et appartenance ethnique en Louisiane francophone." In *La phonologie du français: Normes, périphéries, modélisation. Mélanges pour Chantal Lyche*, edited by Jacques Durand, Gjert Kristoffersen, and Bernard Laks, 289–305. Paris: Presses Universitaires de Paris Ouest.

Klingler, Thomas A. 2015. "Beyond Cajun: Toward an Expanded View of Regional French in Louisiana." In *New Perspectives on Language Variety in the South: Historical and Contemporary Approaches*, edited by Michael D. Picone and Catherine Evans Davies, 627–40. Tuscaloosa: University of Alabama Press.

Klingler, Thomas A. 2019. "The Louisiana Creole Language Today." In *Language in Louisiana: Community and Culture*, edited by Nathalie Dajko and Shana Walton, 90–107. Jackson: University Press of Mississippi.

Klingler, Thomas A., and Nathalie Dajko. 2006. "Louisiana Creole at the Periphery." In *History, Society, and Variation: Studies in Honor of Albert Valdman*, edited by Clancy J. Clements, Thomas A. Klingler, Deborah Piston-Hatlen, and Kevin J. Rottet, 11–28. Amsterdam: Benjamins.

Kniffen, Fred B., Hiram F. Gregory, and George A. Stokes. 1987. *The Historic Indian Tribes of Louisiana: From 1542 to the Present*. 2nd ed. Baton Rouge: Louisiana State University Press.

Labov, William. 1966. *The Social Stratification of English in New York City*. Washington, DC: Center for Applied Linguistics.

Labov, William. 1972. *Sociolinguistic Patterns*. Philadelphia: University of Pennsylvania Press.

Labov, William. 1984. "Field Methods of the Project on Linguistic Change and Variation." In *Language in Use: Readings in Sociolinguistics*, edited by John Baugh and Joel Scherzer, 28–53. Englewood Cliffs: Prentice-Hall.

Lagamayo, Anne. 2017. "Climate Change Threatens to Wash Away Couple's History." CNN online, March 2. http://edition.cnn.com/2017/03/02/us/heart-of-the-matter-climate-change-louisiana/index.html.

Landry, Rodrigue, and Richard Bourhis. 1997. "Linguistic Landscape and Ethnolinguistic Vitality: An Empirical Study." *Journal of Language and Social Psychology* 16 (1): 23–49.

Larouche, Alain. 1981. "Ethnicité, pêche, et petrole: Les cajins du Bayou Lafourche en Louisiane française." Master's thesis, York University.

Lecompte, Nolan Philip, Jr. 1962. "A Word Atlas of Terrebonne Parish." Master's thesis, Louisiana State University.

Lecompte, Nolan Philip, Jr. 1967. "A Word Atlas of Lafourche Parish and Grand Isle, Louisiana." PhD diss., Louisiana State University.

Ledet, Laïse. 1982. *They Came, They Stayed: Origins of Pointe aux Chênes and Ile à Jean Charles, a Genealogical Study, 1575–1982*. http://www.rootsweb.ancestry.com/~laterreb/theycame/index-1.htm.

Levin, Dan. 2016. "Proud to Live in a Town Called Dildo." *New York Times*, July 4.

Lindner, Tamara. 2019. "The Future of French in Louisiana." In *Language in Louisiana: Community and Culture*, edited by Nathalie Dajko and Shana Walton, 108–24. Jackson: University Press of Mississippi.

Lodge, Anthony. 2004. *A Sociolinguistic History of Parisian French*. Cambridge: Cambridge University Press.

Lombard, Gervais. 1944. "Terrebonne Parish Boundaries. Revised as of July 24th, 1944." Report submitted to the Department of Public Works, Baton Rouge, August 7. Copy in possession of author.

Lordveus. 2016. "Hoo Boy." *Trump Explains How To Pronounce "Nevada," Nevadans Correct Him.* https://www.reddit.com/r/politics/comments/562vn9/trump_explains_how_to _pronounce_nevada_nevadans.

Low, Setha M. 1994. "Cultural Conservation of Place." In *Conserving Culture: A New Discourse on Heritage*, edited by Mary Hufford, 66–77. Urbana: University of Illinois Press.

Lyche, Chantal. 1995. "Schwa Metathesis in Cajun French." *Folia Linguistica* 29 (3–4): 369–93.

Maldonado, Julie. 2014. "Facing the Rising Tide: Co-Occurring Disasters, Displacement, and Adaptation in Coastal Louisiana's Tribal Communities." PhD diss., American University.

Marshall, Bob. 2017. "Coastal Flooding May Force Thousands of Homes in Louisiana to Be Elevated or Bought Out." *The Lens*, January 3. http://www.nola.com/environment/index .ssf/2017/04/study_predicts_mass_inland_mig.html.

Marshall, Margaret. 1989. "The Origins of Creole French in Louisiana." *Regional Dimensions* 8:23–40.

Martinez, Paula, David Eads, and Christopher Groskopf. 2015. "Post-Katrina New Orleans Smaller, but Population Growth Rates Back on Track." NPR, August 19. https://www.npr .org/2015/08/19/429353601/post-katrina-new-orleans-smaller-but-population-growth -rates-back-on-track?t=1532738685836.

McQuaid, John, and Mark Schleifstein. 2006. *Path of Destruction: The Devastation of New Orleans and the Coming Age of Superstorms.* New York: Warner.

Menn, Joseph Karl. 1964. *The Large Slave Holders of Louisiana—1860.* New Orleans: Pelican.

Michot, Stephen S. 1996. "Lafourche Society in 1860." In *The Lafourche Country II: The Heritage and Its Keepers*, edited by Stepen S. Michot and John P. Doucet, 26–38. Thibodaux: Lafourche Heritage Society.

Miller, Mark Edwin. 2004. *Forgotten Tribes: Unrecognized Indians and the Federal Acknowledgement Process.* Lincoln: University of Nebraska Press.

Monmonier, Mark. 2006. *From Squaw Tit to Whorehouse Meadow: How Maps Name, Claim, and Inflame.* Chicago: University of Chicago Press.

National Geographic Society. 1930. *Louisiana.* Washington: National Geographic Society. http://digitool.is.cuni.cz:1801/view/action/nmets.do?DOCCHOICE=1176308.xml&dv s=583711031533~333&locale=en_US&search_terms=&adjacency=&VIEWER_URL=/ view/action/nmets.do?&DELIVERY_RULE_ID=3&divType=.

National Map Company. 1927. *Louisiana.* Indianapolis: National Map Company https://www .davidrumsey.com/luna/servlet/detail/RUMSEY~8~1~256938~5520643:Louisiana?sort =pub_list_no_initialsort%2Cpub_date%2Cpub_list_no%2Cseries_no&qvq=w4s:/wher e%2FUnited%2BStates%2FLouisiana;sort:pub_list_no_initialsort%2Cpub_date%2Cpub_ list_no%2Cseries_no;lc:RUMSEY~8~1&mi=22&trs=40.

Neumann-Holzschuh, Ingrid. 1987. *Textes anciens en créole louisianais: Avec introduction, notes, remarques sur la langue et glossaire.* Hamburg: Buske.

New Orleans Daily Picayune. 1893a. "Over a Thousand Lives Lost." October 5, 1.

New Orleans Daily Picayune. 1893b. "The Full Story of Cheniere Isle." October 6, 1.

New Orleans Daily Picayune. 1893c. "Down Bayou Lafourche." October 11, 1.

Nuttall, Mark. 2001. "Locality, Identity and Memory in South Greenland." *Études/Inuit/ Studies* 25 (1–2): 53–72.

Oubre, Elton. 1985. "The European Settlement of the Lafourche Country." In *The Lafourche Country: The People and the Land*, edited by Philip Uzee, 1. Thibodaux: Lafourche Heritage Society.

Oukada, Larbi. 1977. "Louisiana French: A Linguistic Study with a Descriptive Analysis of Lafourche Dialect." PhD diss., Louisiana State University.

Papen, Robert. 2004. "Sur quelques aspects structuraux du français des Métis de l'Ouest Canadien." In *Variation et francophonie*, edited by Aidan Coveney, Marie-Anne Hintze and Carol Sanders, 105–29. Paris: L'Harmattan.

Papen, Robert, and Kevin J. Rottet. 1997. "A Structural Sketch of the Cajun French Spoken in Terrebonne and Lafourche Parishes." In *French and Creole in Louisiana*, edited by Albert Valdman, 71–108. New York: Plenum.

Parenton, Vernon J., and Roland J. Pellegrin. 1950. "The 'Sabines': A Study of Racial Hybrids in a Louisiana Coastal Parish." *Social Forces* 29:148–54.

Parish of Lafourche v. Parish of Terrebonne. 1882. 34 La.Ann. 1230. Supreme Court of Louisiana. No. 8457. December.

Parker, Nancy. 2017. "Echoes of Injustice: The Image of a Slave Brings Closure to a Terrebonne Parish Family." Fox8 News, May 24. http://www.fox8live.com/story/35502367/ echoes-of-injustice-the-image-of-a-slave-brings-closure-to-a-terrebonne-parish-family.

Parr, Una M. 1940. "A Glossary of the Variants from Standard French in Terrebonne Parish: With an Appendix of Popular Beliefs, Superstitions, Medicines and Cooking Recipes." Master's thesis, Louisiana State University.

Picone, Michael D. 1994. "Code-Intermediate Phenomena in Louisiana French." In *CLS 30-I: Papers from the Thirtieth Regional Meeting of the Chicago Linguistic Society*, vol. 1, *The Main Session*, edited by Katie Beals, Jeannette Denton, Bob Knippen, Lynette Melnar, Hisami Suzuki, and Erika Zeinfeld, 320–34. Chicago: Chicago Linguistic Society.

Picone, Michael D. 1997a. "Code-Switching and Loss of Inflection in Louisiana French." In *Language Variety in the South Revisited*, edited by Cynthia Bernstein, Thomas E. Nunnally, and Robin Sabino, 152–62. Tuscaloosa: University of Alabama Press.

Picone, Michael D. 1997b. "Enclave Dialect Contraction: An External Overview of Louisiana French." *American Speech* 72 (2): 117–53.

Picone, Michael D. 1998. "Historic French Diglossia in Louisiana." Paper presented at the Annual Meeting of the Southeastern Conference on Linguistics, University of Southwestern Louisiana, Lafayette, March 26–28.

Picone, Michael D. 2003. "Anglophone Slaves in Francophone Louisiana." *American Speech* 78(4):404–433.

Picone, Michael D. 2006. "Le français hors de l'Acadiana." *Revue de l'Université de Moncton* 37:221–31.

Picone, Michael D. 2014. "Literary Dialect and the Linguistic Reconstruction of Nineteenth-Century Louisiana." *American Speech* 89 (2): 143–69.

Picone, Michael D. 2015. "French Dialects of Louisiana: A Revised Typology." In *New Perspectives on Language Variety in the South: Historical and Contemporary Approaches*, edited by Michael D. Picone and Catherine Evans Davies, 267–87. Tuscaloosa: University of Alabama Press.

Picone, Michael D., and Albert Valdman. 2005. "La situation du français en Louisiane." In *Le français en Amérique du Nord*, edited by Albert Valdman, Julie Auger, and Deborah Piston-Hatlen, 143–65. Quebec City: Les Presses de l'Université Laval.

Pierron, Walter Joseph. 1942. "Sociological Study of the French-Speaking People in Chauvin." Master's thesis, University of Louisiana.

Pitre, Loulan J., Jr. 1996. "Chenière Caminada, A Late Nineteenth-Century Coastal Cajun Community." In *The Lafourche Country II: The Heritage and Its Keepers*, edited by Stephen Michot and John Doucet, 51–70. Thibodaux: Lafourche Heritage Society.

Poirier, Pascal. 1928. *Le parler Franco-Acadien et ses origines*. Quebec City: Imprimerie franciscaine missionaire.

Police Jury of Parish of Lafourche v. Police Jury of Parish of Terrebonne. 1896. 48 La.Ann.1299. Supreme Court of Louisiana. No. 12,114. June 1.

Police Jury of Parish of Lafourche v. Police Jury of Parish of Terrebonne. 1897. 49 La.Ann.1331. Supreme Court of Louisiana. No. 12,468. May 31.

Pointe au Chien Indian Tribe. N.d. "History/Background." http://pactribe.tripod.com/id2.html.

Pointe au Chien Indian Tribe. 2005. *Petition for Federal Recognition, part II*. Montegut: Pointe au Chien Indian Tribe.

Pointe au Chien Indian Tribe. 2012. "Pointe au Chien Indian Tribe Holds First Culture Camp." Press release, July 20. http://pactribe.tripod.com/id22.html.

Preston, Dennis. 1986. "Five Visions of America." *Language in Society* 15 (2): 221–40.

Preston, Dennis. 1996. "Where the Worst English Is Spoken." In *Focus on the USA*, edited by Edgar Schneider, 297–360. Amsterdam: Benjamins.

Preston, Dennis. 1999. *Handbook of Perceptual Dialectology*. Amsterdam: Benjamins.

Preston, Dennis. 2011. "Methods in (Applied) Folk Linguistics: Getting into the Minds of the Folk." *AILA Review* 24:5–39.

Rand McNally. 1889. "Rand, McNally & Co.'s Louisiana." *Rand McNally & Co.'s Enlarged Business and Shippers Guide . . . Together with a Complete Reference Map of the World*. Chicago: Rand McNally. https://www.davidrumsey.com/luna/servlet/detail/RUMSEY~8 ~1~37429~1210303:Louisiana-?sort=pub_list_no_initialsort%2Cpub_date%2Cpub_list_ no%2Cseries_no&qvq=w4s:/where%2FUnited%2BStates%2FLouisiana;sort:pub_list_no_ initialsort%2Cpub_date%2Cpub_list_no%2Cseries_no;lc:RUMSEY~8~1&mi=10&trs=40.

Rand McNally. 1939. *Rand McNally Road Map: Arkansas—Louisiana—Mississippi*. Bloomington: State Farm Insurance Companies Travel Bureau. https://www.david rumsey.com/luna/servlet/detail/RUMSEY~8~1~258596~5522092:Rand-McNally -Road-map--Arkansas,-Lo?sort=pub_list_no_initialsort%2Cpub_date%2Cpub_list_ no%2Cseries_no&qvq=w4s:/where%2FUnited%2BStates%2FLouisiana;sort:pub_list_no_ initialsort%2Cpub_date%2Cpub_list_no%2Cseries_no;lc:RUMSEY~8~1&mi=38&trs=40.

Ratliff, Robert Barrow. Notebook. Barrow Family Papers, box 20, folder 2. Manuscripts Collection 654, Howard Tilton Memorial Library, Tulane University, New Orleans.

Read, William A. (1931) 1963. *Louisiana-French.* Baton Rouge: Louisiana State University Press.

Rhodes, Richard A. 2009. "The Phonological History of Métchif." In *Le français d'un continent à l'autre: Mélanges offertes à Yves Charles Morin,* edited by Luc Baronian and France Martineau, 423–42. Quebec City: Presses de l'Université de Laval.

Rootsweb. n.d. *Terrebonne Parish Map of 1909.* http://sites.rootsweb.com/~laterreb/1909map .htm.

Rottet, Kevin J. 1995. "Language Shift and Language Death in the Cajun French-Speaking Communities of Terrebonne and Lafourche Parishes, Louisiana." PhD diss., Indiana University.

Rottet, Kevin J. 2001. *Language Shift in the Coastal Marshes of Louisiana.* New York: Lang.

Rottet, Kevin J. 2004. "Inanimate Interrogatives and Settlement Patterns in Francophone Louisiana." *French Language Studies* 14:169–88.

Rousseau, Jean. 1852. "Partis Marrons." *Le pionnier de l'Assomption,* March 18, 3. Available at Louisiana Runaway Slave Advertisements, 1836–1865, https://louisianadigitallibrary.org/ islandora/object/lsu-sc-p16313coll80%3A1911.

Salmon, Carole Lucienne. 2007. "Français acadien, français cadien: Variation stylistique et maintenance de formes phonétiques dans le parler de quatre générations de femmes cadiennes." PhD diss., Louisiana State University.

Schafer, Mark. 1993. *The Soundscape: Our Sonic Environment and the Tuning of the World.* Rochester: Destiny Books.

Schladebeck, Jessica. 2017. "Louisiana Island That's Sinking Underwater Is Home to First 'Climate Change Refugees.'" *New York Daily News,* March 2.

Schreyer, Christine. 2008. "Taku River Tlingit Genres of Place as Performatives of Stewardship." *Journal of Linguistic Anthropology* 26 (1): 4–25.

Scott, Ann Martin. 1992. "Some Phonological and Syntactic Characteristics of Cajun Vernacular English." In *Cajun Vernacular English: Informal English in French Louisiana,* edited by Ann Martin Scott, 26–37. Lafayette: University of Southwestern Louisiana Press.

Sebba, Marc. 2010. "Discourses in Transit." In *Semiotic Landscapes: Language, Image, Space,* edited by Adam Jaworski and Crispin Thurlow, 59–76. London: Continuum International.

Shepherd, Nick. 2001. "Comments on Part II: Far from Home." In *Contested Landscapes: Movement, Exile, and Place,* edited by Barbara Bender and Margot Winer, 349–58. Oxford: Berg.

Shohamy, Elana, and Shoshi Waksman. 2009. "Linguistic Landscape as an Ecological Arena: Modalities, Meanings, Negotiations, Education." In *Linguistic Landscape: Expanding the Scenery,* edited by Elana Shohamy and Durk Gorter, 313–31. New York: Routledge.

Sibata, Takesi. (1959) 1999. "Consciousness of Dialect Boundaries." Translated by Daniel Long. In *Handbook of Perceptual Dialectology,* edited by Dennis Preston, 1:39–62. Amsterdam: Benjamins.

Silverstein, Michael. 2014. "The Race from Place: Dialect Eradication vs. the Linguistic 'Authenticity' of Terroir." In *Indexing Authenticity: Sociolinguistic Perspectives,* edited by Veronique Lacoste, Jakob Leimgruber, and Thiemo Breyer, 159–87. Boston: de Gruyter.

Speck, Frank. 1943. "A Social Reconnaissance of the Creole Houma Trappers of the Louisiana Bayous." *America Indigena* 3:134–46, 211–20.

Stanton, Max E. 1971. "A Remnant Indian Community: The Houma of Southern Louisiana." In *The Not-So-Solid South: Anthropological Studies in a Regional Subculture*, edited by John Kenneth Morland, 82–92. Athens: University of Georgia Press.

Stanton, Max E. 1979. "Southern Louisiana Survivors: The Houma Indians." In *Southeastern Indians since the Removal Era*, edited by Walter L. Williams, 90–109. Athens: University of Georgia Press.

State of Emergency—Coastal Louisiana. 2017. Executive Department Proclamation 43 JBE 2017, April 18. http://gov.louisiana.gov/assets/EmergencyProclamations/43-JBE-2017-Coastal-Louisiana.pdf.

Swanton, John Reed. 1907. "Southeast ethnographic and vocabulary notes." National Anthropological Archives, MS 4201. Suitland, MD: Smithsonian Institution.

Swanton, John R. 1911. *Indian Tribes of the Lower Mississippi Valley and Adjacent Coast of the Gulf of Mexico*. Washington, DC: Smithsonian Institution.

Terrebonne Genealogical Society. 2019. Conveyance Records Acts 201–400. Transcribed by Phil Chauvin Jr. http://www.terrebonnegenealogicalsociety.org/conveyance-records-acts-201-400/.

Thomas, Gilles. 2018. Personal communication, July 23.

Thurston, Dee Dee. 2002. "Pointe-aux-Chenes Wouldn't Be the Same without Laise Ledet." *Houma Courier*, April 9. http://www.rootsweb.ancestry.com/~laterreb/theycame/images/obitlais.htm.

Tidwell, Mike. 2004. *Bayou Farewell: The Rich Life and Tragic Death of Louisiana's Cajun Coast*. New York: Pantheon.

Tidwell, Mike. 2006. *The Ravaging Tide: Strange Weather, Future Katrinas, and the Coming Death of America's Coastal Cities*. New York: Free Press.

Trépagnier, Numa. 1855. "Detained in the Jail of the Parish of St. John the Baptist." *Le Meschacébé*, July 1, 2. Available at Louisiana Runaway Slave Advertisements, 1836–1865, https://louisianadigitallibrary.org/islandora/object/lsu-sc-p16313coll80%3A420.

Trépanier, Cécyle. 1991. "The Cajunization of French Louisiana: Forging a Regional Identity." *Geographical Journal* 125:161–71.

Trudgill, Peter. (1985). "New Dialect Formation and the Analysis of Colonial Dialects: The Case of Canadian Raising." In *Papers from the Fifth International Conference on Methods in Dialectology*, ed. H. J. Warkentyne, 35–45. Victoria, BC: University of Victoria.

Trudgill, Peter. 1986. *Dialects in Contact*. Oxford: Blackwell

Tuan, Yi-Fu. 1975. "Place: An Experiential Perspective." *Geographical Review* 65 (2): 151–65.

Tuan, Yi-Fu. 1991. "Language and the Creation of Place: A Narrative-Descriptive Approach." *Annals of the Association of American Geographers* 81 (4): 684–96.

Tuck, Eve, and Marcia McKenzie. 2015. *Place in Research: Theory, Methodology, and Methods*. New York: Routledge.

United Houma Nation. N.d. "Federal Recognition Timeline." http://www.unitedhoumanation.org/node/770.

United Nations High Commissioner for Refugees. 2019. "Figures at a Glance." Last updated June 19. http://www.unhcr.org/figures-at-a-glance.html.

US Bureau of Indian Affairs. 1994. "Summary under the Criteria and Evidence for Proposed Finding against Federal Acknowledgment of the United Houma Nation, Inc." https://www.bia.gov/sites/bia.gov/files/assets/as-ia/ofa/petition/056_uhouma_LA/056_pf.pdf.

US Bureau of Indian Affairs. 2008a. "Amended Proposed Finding against Acknowledgment of the Pointe-au-Chien Indian Tribe (PACIT) of Louisiana." *Federal Register* document 73 FR 31142. https://www.federalregister.gov/documents/2008/05/30/E8-12153/amended-proposed-finding-against-acknowledgment-of-the-pointe-au-chien-indian-tribe-pacit-of.

US Bureau of Indian Affairs. 2008b. "Amended Proposed Finding against Acknowledgment of the Biloxi, Chitimacha Confederation of Muskogees, Inc (BCCM) of Louisiana." *Federal Register* document 73 FR 31140. https://www.federalregister.gov/documents/2008/05/30/E8-12155/amended-proposed-finding-against-acknowledgment-of-the-biloxi-chitimacha-confederation-of-muskogees

US Census Bureau. 1850a. "1850 U. S. Census." *Seventh Census of the United States, 1850.* National Archives Microfilm Publication M432, 1009 rolls. Records of the Bureau of the Census, Record Group 29, National Archives, Washington, D.C.

US Census Bureau. 1850b. "1850 U.S. Federal Census—Slave Schedules." *Seventh Census of the United States, 1850.* National Archives Microfilm Publication M432, 1,009 rolls. Records of the Bureau of the Census, Record Group 29, National Archives, Washington, D.C.

US Census Bureau. 1860a. 1860 U.S. Federal Census—Slave Schedules. *Eighth Census of the United States, 1860.* Washington, D.C.: National Archives and Records Administration, 1860. M653, 1,438 rolls.

US Census Bureau. 1860b. *Table No. 2: Population by Color and Condition*, p. 194. https://www2.census.gov/library/publications/decennial/1860/population/1860a-16.pdf.

US Census Bureau. 2000. "Language Spoken at Home for Counties and Tracts in Louisiana: 2000." Special Tabulation 224. Excel file downloaded from www.census.gov, December 13, 2006.

US Census Bureau. 2010a. "Race and Hispanic or Latino Origin: 2010." Summary File 1. https://factfinder.census.gov/faces/tableservices/jsf/pages/productview.xhtml?src=CF.

US Census Bureau. 2010b. *Place of Birth for the Foreign-Born Population. Universe: Orleans, Jefferson, St. Bernard Parishes, 2010.* https://factfinder.census.gov/faces/tableservices/jsf/pages/productview.xhtml?pid=ACS_10_5YR_B05006&prodType=table.

US Census Bureau. 2013. *Detailed Languages Spoken at Home and Ability to Speak English for the Population 5 Years and Over: 2009–2013.* https://www.census.gov/data/tables/2013/demo/2009-2013-lang-tables.html.

US Census Bureau. 2016. *Comparative Demographic Estimates, 2012–2016 American Community Survey 5-Year Estimates.* https://factfinder.census.gov/faces/tableservices/jsf/pages/productview.xhtml?pid=ACS_16_5YR_CP05&prodType=table.

US Census Bureau. 2017. *Place of Birth for the Foreign-Born Population. Universe: Orleans, Jefferson, St. Bernard Parishes, 2017.* https://factfinder.census.gov/faces/tableservices/jsf/pages/productview.xhtml?pid=ACS_17_5YR_B05006&prodType=table.

US Geological Survey. 2017a. "Dog." Geographic Names Information System (GNIS). Search results retrieved June 19. https://geonames.usgs.gov/apex/f?p=138:2:0::NO:RP::.

US Geological Survey. 2017b. "Oak." Geographic Names Information System (GNIS). Search results retrieved June 19. https://geonames.usgs.gov/apex/f?p=138:2:0::NO:RP::.

US Geological Survey. 2019a. "Bayou de Chien." Geographic Names Information System (GNIS). Search results retrieved September 22. https://geonames.usgs.gov/apex/f?p=138:3:0::NO:3:P3_FID,P3_TITLE:486489,Bayou%20de%20Chien.

US Geological Survey. 2019b. "River aux Chenes." Geographic Names Information System (GNIS). Search results retrieved August 2. https://geonames.usgs.gov/apex/f?p=138:3:0::NO:3:P3_FID,P3_TITLE:558302,River%20aux%20Chenes.

Valdman, Albert. 2007. "Vernacular French Communities in the United States: A General Survey." *French Review* 80:1218–34.

Valdman, Albert, Kevin J. Rottet, Barry Jean Ancelet, Richard Guidry, Amanda LaFleur, Thomas A. Klingler, Tamara Lindner, Michael D. Picone, and Dominique Ryon. 2010. *Dictionary of Louisiana French: As Spoken in Cajun, Creole, and Native American Communities.* Jackson: University Press of Mississippi.

Van De Bogart, Betty. 2003. "Letter to the Editor." *Reno Gazette*, December 1, 6A.

Van Houten, Carolyn. 2016. "The First Official Climate Refugees in the U.S. Race against Time." *National Geographic News*, May 25. http://news.nationalgeographic.com/2016/05/160525-isle-de-jean-charles-louisiana-sinking-climate-change-refugees/

Vaux, Bert, and Scott Golder. 2003. "Caramel." *The Harvard Dialect Survey.* https://www4.uwm.edu/FLL/linguistics/dialect/staticmaps/q_4.html.

Walker, Douglas C. 1984. *The Pronunciation of Canadian French.* Ottawa: University of Ottawa Press.

Walton, Shana. 1994. "Flat Speech and Cajun Ethnic Identity in Terrebonne Parish, Louisiana." PhD diss., Tulane University.

Walton, Shana. 2017a. Personal communication, May 2.

Walton, Shana. 2017b. Personal communication, December 19.

West, Robert C. 1986. *An Atlas of Surnames of French and Spanish Origin.* Baton Rouge: Louisiana State University Press.

Westerman, Audrey. N.d. *Preliminary Genealogical and Historical Information of the Biloxi-Chitimacha Confederation of Muskogees, Inc.* Houma: Biloxi-Chitimacha Confederation of Muskogees.

Wolff, Hans. 1959. "Intelligibility and Inter-Ethnic Attitudes." *Anthropological Linguistics* 1 (3): 34–41.

Zinsli, Paul. 1957. *Berndeutsche Mundart* [Bern German dialect]. Bern: Berner Staatsbuch.

INDEX

References to illustrations and tables appear in **bold**.

ABOUT THE AUTHOR

Nathalie Dajko is associate professor of anthropology at Tulane University, New Orleans, where she studies Louisiana's French and English varieties. She has conducted hundreds of interviews with residents of South Louisiana and has done years of participant observation in both rural and urban settings. She has published in the *Journal of Linguistic Anthropology*, *Language in Society*, and in several edited volumes in both French and English. She is coeditor of *Language in Louisiana: Community and Culture*, also published by the University Press of Mississippi.

CPSIA information can be obtained
at www.ICGtesting.com
Printed in the USA
BVHW030259211020
591446BV00001B/3